CONTENTS

Letter of Transmittal to the President v
Letter from the President to the Commission vi
Members of the Commission vii
Commission Staff and Consultants viii
Acknowledgements ix
Executive Summary 1
    Improving Accountability 6
    Treating and Compensating for Research-Related Injury 8
    Creating a Culture of Responsibility: Human Research Protections as
    Professional Standards 9
    Respecting Equivalent Protections 10
    Promoting Community Engagement 11
    Justifying Site Selection 12
    Ensuring Ethical Study Design 13
    Promoting Current Federal Reform Efforts 14
    Following Up 15
CHAPTER 1: Introduction 17
    The Need to Assess the Contemporary Environment 21
    Research Across Borders 23
    Contemporaneous Reviews of the Human Subjects Protection System 25
    About this Report 26
CHAPTER 2: Assessing the Current System 29
    The Scope and Volume of Federal Human Subjects Research 33
    Current Rules 40
    Conclusion 42
CHAPTER 3: Further Analysis and Recommendations 43
    Commission Recommendations 46
        1. Improving Accountability 46
        2. Treating and Compensating for Research-Related Injury 56
            Ethical Justification for Compensation 56
            Reparations for Unethical Research 62
            Designing a System of Compensation 64

3. Creating a Culture of Responsibility: Human Research Protections as Professional Standards 70
4. Respecting Equivalent Protections... 74
5. Promoting Community Engagement 78
6. Justifying Site Selection... 82
7. Ensuring Ethical Study Design 88
    Towards a Middle Ground on Study Design 91
        (1) Treatment Standards 91
        (2) Methodological Constraints... 93
        (3) Risk Minimization 94
8. Promoting Current Reform Efforts 96
    Ensuring Risk-Based Protections.. 96
    Streamlining IRB Review of Multi-Site Studies 98
    Improving Informed Consent 99
    Clarifying and Harmonizing Regulatory Requirements and Agency Guidance... 100
    Data Collection to Enhance Adverse Event Reporting 100
9. Following Up 102
Endnotes 105
Appendices... 139
    Appendix I: Human Subjects Research Landscape Project: Scope and Volume of Federally Supported Human Subjects Research 140
    Appendix II: Human Subjects Research Landscape Project Methods............ 163
    Appendix III: U.S. Treatment/Compensation for Treatment Methods.......... 184
    Appendix IV: International and Transnational Requirements for Treatment and Compensation for Research Injuries...................................... 186
    Appendix V: International Research Panel 191
    Appendix VI: Guest Speakers 192

Presidential Commission for the Study of Bioethical Issues

President Barack Obama
The White House
1600 Pennsylvania Avenue, NW
Washington, DC 20500

Dear Mr. President:

On behalf of the Presidential Commission for the Study of Bioethical Issues, we present to you this report, *"Moral Science: Protecting Participants in Human Subjects Research."*

In response to your request of November 24, 2010, the Commission oversaw a thorough review of current regulations and international standards to assess whether they adequately protect human participants in federally funded research, no matter where it occurs.

The Commission held four multi-day, public meetings. Speakers addressed a range of U.S. and international policies and norms, and provided perspectives from a wide array of professional and institutional organizations. At your request, we sought the advice of international experts and appointed the International Research Panel, a subcommittee of the Commission. Finally, the Commission solicited information from the public and it received over three hundred comments.

The Commission concludes that current regulations, which apply to a diverse and wide-ranging portfolio of research, generally appear to protect people from avoidable harm or unethical treatment. However, because of the currently limited ability of some governmental agencies to identify basic information about all of their human subjects research, the Commission cannot say that all federally funded research provides optimal protections against avoidable harms and unethical treatment.

Many of our most important advances derive from research that involves human participants. It is essential, therefore, that critical research of this sort adheres to the highest ethical practices to ensure, as best as possible, that those who volunteer to participate in studies for the benefit of others are protected. Thus, the Commission offers 14 recommendations to improve the current system.

The Commission Members are honored by the trust you have placed in us and we are grateful for the opportunity to serve you and the nation in this way.

Sincerely,

Amy Gutmann, Ph.D.
Chair

James Wagner, Ph.D.
Vice-Chair

1425 New York Avenue, NW, Suite C-100, Washington, DC 20005
Phone 202-233-3960 Fax 202-233-3990 www.bioethics.gov

THE WHITE HOUSE
WASHINGTON

November 24, 2010

MEMORANDUM FOR DR. AMY GUTMANN
                Chair, Presidential Commission for the Study of
                Bioethical Issues

SUBJECT: Review of Human Subjects Protection

Recently, we discovered that the U.S. Public Health Service conducted research on sexually transmitted diseases in Guatemala from 1946 to 1948 involving the intentional infection of vulnerable human populations. The research was clearly unethical. In light of this revelation, I want to be assured that current rules for research participants protect people from harm or unethical treatment, domestically as well as internationally.

I ask you, as the Chair of the Presidential Commission for the Study of Bioethical Issues, to convene a panel to conduct, beginning in January 2011, a thorough review of human subjects protection to determine if Federal regulations and international standards adequately guard the health and well-being of participants in scientific studies supported by the Federal Government. I also request that the Commission oversee a thorough fact-finding investigation into the specifics of the U.S. Public Health Service Sexually Transmitted Diseases Inoculation Study.

In fulfilling this charge, the Commission should seek the insights and perspective of international experts, including from Guatemala; consult with its counterparts in the global community; and convene at least one meeting outside the United States. I expect the Commission to complete its work within 9 months and provide me with a report of its findings and recommendations.

While I believe the research community has made tremendous progress in the area of human subjects protection, what took place in Guatemala is a sobering reminder of past abuses. It is especially important for the Commission to use its vast expertise spanning the fields of science, policy, ethics, and religious values to carry out this mission. We owe it to the people of Guatemala and future generations of volunteers who participate in medical research.

PRESIDENTIAL COMMISSION FOR THE STUDY OF BIOETHICAL ISSUES

AMY GUTMANN, PH.D., CHAIR
President and Christopher H. Browne
Distinguished Professor of Political Science, University of Pennsylvania

JAMES W. WAGNER, PH.D., VICE CHAIR
President, Emory University

YOLANDA ALI, M.B.A.
Michael J. Fox Foundation
Founder's Council;
Emory Neurosciences
Community Advisory Board

ANITA L. ALLEN, J.D., PH.D.
Henry R. Silverman Professor of Law
And Professor of Philosophy
University of Pennsylvania Law School

JOHN D. ARRAS, PH.D.
Porterfield Professor of Biomedical
Ethics, Professor of Philosophy,
University of Virginia

BARBARA F. ATKINSON, M.D.
Executive Vice Chancellor,
University of Kansas Medical Center;
Executive Dean, University of
Kansas School of Medicine

NITA A. FARAHANY, J.D., PH.D.
Leah Kaplan Visiting Professor in
Human Rights, Stanford Law School
Associate Professor of Law;
Associate Professor of Philosophy
Vanderbilt University

ALEXANDER G. GARZA, M.D., M.P.H.
Assistant Secretary, Office of Health
Affairs; Chief Medical Officer,
Department of Homeland Security

CHRISTINE GRADY, R.N., PH.D.
Acting Chief of the Department of
Bioethics, National Institutes of
Health Clinical Center

STEPHEN L. HAUSER, M.D.
Robert A. Fishman Distinguished
Professor and Chair of the Department
of Neurology, University of California,
San Francisco

RAJU S. KUCHERLAPATI, PH.D.
Paul C. Cabot Professor, Department
of Genetics, Harvard Medical School;
Professor, Department of Medicine,
Brigham and Women's Hospital

NELSON L. MICHAEL, M.D., PH.D.
Colonel, Medical Corps, U.S. Army;
Director, Division of Retrovirology;
Walter Reed Army Institute of Research;
U.S. Military HIV Research Program

DANIEL P. SULMASY, M.D., PH.D, FACP
Kilbride-Clinton Professor of Medicine
and Ethics, Department of Medicine
and Divinity School; Associate Director,
The MacLean Center for Clinical Medical
Ethics, University of Chicago

PRESIDENTIAL COMMISSION FOR THE STUDY OF BIOETHICAL ISSUES

STAFF* AND CONSULTANTS

Executive Director
Valerie H. Bonham, J.D.

Deputy Director
Debbie Banks Forrest, M.P.P.

Communications Director
Hillary Wicai Viers, M.S.J.

Senior Advisors
Paul Lombardo, Ph.D., J.D.
Jonathan D. Moreno, Ph.D.
Jeremy Sugarman, M.D., M.P.H., M.A.

Research Staff
Eleanor Celeste, B.A.
Tom Cinq-Mars, B.A.
Brian C. Eiler, J.D.
Michelle Groman, J.D.
Chris Havasy, Sc.B.
Holly Fernandez Lynch, J.D.,
 M. Bioethics
Debra Mathews, Ph.D.
Eleanor E. Mayer, J.D., M. Bioethics
Anne Pierson, J.D.
Elizabeth Pike, J.D.
Cary Scheiderer, Ph.D
Kayte Spector-Bagdady, J.D.,
 M. Bioethics
Victoria Wilbur, B.A.

Consultants
Burness Communications
Kathi E. Hanna, M.S., Ph.D.

Committee and Staff Affairs
Svetlana Cicale, M.S.
Judith E. Crawford
Esther E. Yoo, B.A.

Fellows and Interns
Adebukola J. Awosogba, M.A.
Michael Grippaldi, B.S.
Magdalina Gugucheva, B.A.
Casey Nicol, B.S.
Tuua Ruutiainen, B.A.
Michael Tennison, M.A.
David Tester, Ph.D.
Jennifer Therrien, B.S.
Ilana Yurkiewicz, B.S.

Includes former and part-time staff

ACKNOWLEDGEMENTS

The Commission is grateful to the many individuals who contributed their time, energy, and expertise to this report. The Commission offers its particular thanks to the exceptional speakers who travelled great distances to participate in the Commission's public meetings. It thanks also the dedicated department and agency employees who shared their experience, expertise, and insight with the Commission.

The Commission also extends its gratitude to the members of the subcommittees whose advice shaped the direction of this report. We thank the members of the International Research Panel for their reflections and unique knowledge of human subjects protection standards and practices in international research, and the members of the Empirical Advisory Group, who advised the Commission on how to analyze and interpret data on the scope of scientific studies currently supported by the federal government.

Finally, the Commission is enormously indebted to the members of its talented staff for their dedicated research, and thoughtful insights on human subjects protection in a short period of time. The Commission extends special thanks to Executive Director Valerie Bonham, and Senior Advisors Jonathan Moreno and Jeremy Sugarman. Furthermore, the Commission is particularly grateful to staff lead Michelle Groman for her tireless and extraordinary efforts on its behalf to ensure a comprehensive assessment of this important topic.

MORAL SCIENCE Protecting Participants in Human Subjects Research

EXECUTIVE SUMMARY

Human research serves to ensure the safety of new medicines, establish tolerable exposure levels for environmental and workplace hazards, and determine the effectiveness of new interventions in public health, education, and countless other fields. Without volunteers, these studies would be impossible to conduct. Recognizing society's responsibility to protect human subjects of research from avoidable harm and unethical treatment, President Barack Obama asked the Presidential Commission for the Study of Bioethical Issues (the Commission) to conduct a thorough review of current regulations and international standards to assess whether they adequately protect human subjects in federally supported scientific studies, no matter where they occur.

The Commission's review confirmed that the federal government supports a diverse and wide-ranging portfolio of research, which includes activities funded directly, or by award or sub-award, throughout the world. Support for medical and public health research predominates, but the federal government also supports a large volume of human subjects research in other fields, including social and behavioral sciences and education.

Sound scientific experimentation is rooted in uncertainty and volunteers cannot be immunized from all physical or psychological risks. However, in order for research with human beings to be ethical, human subjects must be volunteers who give their informed consent, who are treated fairly and respectfully, who are subjected only to reasonable risks from which proportionate humanitarian benefit can be obtained, and who are not treated as mere means to the ends of others. (Some carefully specified and regulated exceptions to informed consent are based on the incapacity of some subjects or the very low risk of the experiments, provided that all the other conditions—including fair and respectful treatment—hold.)[1]

In the absence of these ethical constraints, tragic results may follow. Many prior abuses of human subjects are now carefully documented, and some informed the development of today's federal human subjects protection system. Eighteen federal departments and agencies require adherence to a uniform regulatory floor for human subjects research, known as the "Common Rule," which generally requires informed consent, independent ethical review, and the minimization of avoidable risks. These standards apply to all research funded by these departments and agencies, regardless of where

EXECUTIVE SUMMARY

it occurs. The Food and Drug Administration applies essentially the same standards to all research conducted in support of seeking U.S. marketing approval for drugs, devices, and biologics; regardless of the source of funding.

These rules reflect widely accepted principles of ethics. These principles are rooted in longstanding values that find expression in many sources of moral philosophy; theological traditions; and codes, regulations, and rules. They are the bulwark of ethically sound science, or "moral science," as the Commission terms it. Each generation may re-examine how these principles are contextually applied and understood. And, their application or implementation may vary depending on the level of risk that a subject faces. Medical research that poses risk of physical injury rightly raises more concerns than does routine social survey research, for example. Nonetheless, the same ethical principles govern all of these activities, and serve as enduring guideposts that must not be ignored.

The public has a right to expect researchers to abide by rules that satisfy these principles. Researchers themselves benefit from public confidence when they conform to these rules; and with public esteem they earn the ability to conduct potentially important research with public support. Without such earned confidence, research participation may be threatened and critical research jeopardized. More than these measurable effects, society risks irretrievably losing sight of what is inherently owed to fellow human beings and those who deserve special protection by virtue of their willingness to participate in experiments designed to benefit others and advance scientific and social progress.

From time to time society revisits the rules applied to research with human subjects and the implementation of guiding ethical principles. The need for reassessment may arise from challenges presented by novel scientific advances, a perceived mismatch between ethical principles and their implementation, or revelations of abuse. When President Obama charged the Commission with undertaking this review of contemporary human subjects protection standards, he recognized the sacred trust and responsibility that we as a society have to ensure that human research subjects are protected from harm and unethical treatment.[2] The immediate catalyst to action came from newly discovered evidence of unethical activities by U.S. scientists in Guatemala in the late 1940s. The Commission's findings and ethical assessment of those events, documented in its report *"Ethically Impossible" STD Research in Guatemala 1946 to 1948*, illustrate how the quest for scientific knowledge without regard to relevant ethical standards can blind researchers to the humanity of the people they enlist into research.[3]

For this review, the first of its kind by a national bioethics commission in a decade, the President asked the Commission to complete its work in nine months. Recognizing the increasing involvement of foreign sites and partners in human subjects research, the Commission organized a panel of international experts, the International Research Panel (the Panel), consisting of experts in bioethics and biomedical research from 10 countries: Argentina, Belgium, Brazil, China, Egypt, Guatemala, India, Russia, Uganda, and the United States.[4] The Panel, led by Commission Chair Amy Gutmann, held three day-long meetings to discuss research standards and practices around the globe. In their discussions, Panel members drew upon their individual expertise and decades of experience conducting research and developing policy to protect human subjects.

In attempting to assess the current depth and breadth of the federally funded human research enterprise, the Commission quickly learned there is no ready source that comprehensively describes its basic characteristics, such as level of funding, or number of studies, subjects, or geographic locations. Instead, what exists are isolated pockets of information and some descriptive summaries. To better understand, and enable the public to know the scope and volume of "scientific studies supported by, the Federal Government," the Commission therefore asked each Common Rule agency to provide limited, project-specific information on human subjects research supported in Fiscal Year 2010, and to identify trends, if possible, the same information back through Fiscal Year 2006.

The Commission collected basic, project-level data about human subjects research, including study title, number and location of sites, number of subjects, and funding information. These data were compiled into the Commission's "Research Project Database," and analyzed as part of its Human Subjects Research Landscape Project (see further discussion below and Appendices I and II). Among other things, the Commission learned that

EXECUTIVE SUMMARY

the federal government supported more than 55,000 human subjects research projects around the globe in Fiscal Year 2010, mostly in medical and health-related research, but also in other fields such as education and engineering. The Commission also learned that many federal departments and agencies have no ready means to identify basic information about the research they support (e.g., location of study sites) or link funding information with study-level data.

The Commission convened an Empirical Advisory Group, comprised of experts in bioethics, statistics, clinical trial management, and qualitative research, to assist in analysis and interpretation of the Human Subjects Research Landscape Project and suggest future empirical work that could be conducted to evaluate the current human subjects protection system.

In sum, the Commission concludes that current regulations generally appear to protect people from avoidable harm or unethical treatment, insofar as is feasible given limited resources,[5] no matter where U.S.-supported research occurs. This conclusion is fully consistent with, and also qualified by, the large yet incomplete set of information made available to the Commission in the time available to carry out its charge. Specifically, the Commission found:

The current U.S. system provides substantial protections for the health, rights, and welfare of research subjects and, in general, serves to "protect people from harm or unethical treatment" when they volunteer to participate as subjects in scientific studies supported by the federal government. However, because of the currently limited ability of some governmental agencies to identify basic information about all of their human subjects research, the Commission cannot conclude that all federally funded research provides optimal protections against avoidable harms and unethical treatment. The Commission finds significant room for improvement in several areas where, for example, immediate changes can be made to increase accountability and thereby reduce the likelihood of harm or unethical treatment.

In reaching this conclusion, the Commission believes that the ethical principles for human subjects research should not—indeed *must* not—vary depending on the source of funding or location of the research.[6] While the specific methods of implementing the ethical principles of human subjects research are likely to differ, the principles should not. Ethical principles provide the foundation for the rules and regulations that govern human subjects research as well as lay the groundwork upon which everyone who conducts human subjects research must stand.

There is no way to eradicate all risk of harm, particularly in some types of medical and translational research, but the Commission found several important areas where improvement or refinement of the current system is both possible and desirable. It offers guidance and reflection in eight specific areas, all ripe for action or advancement now. Chief among these, the Commission finds that accountability can and should be refined through improving access to basic information about the scope and volume of human subjects research funded by the government. It also draws a bright line affirming the view of most bioethicists and others, including the majority of nations supporting human subjects research around the globe, that human subjects should not individually bear the costs of care required to treat harms resulting directly from that research. The Commission also calls on the federal government to respect the equivalent protections offered by international partners and exercise its longstanding authority to recognize these protections when available.

The Commission's review of the current system comes while the government is already considering systematic reform through revision of the Common Rule. Some of the proposed reforms offer useful means to improve upon current practices. Although the Commission was not asked to undertake a comprehensive assessment of the proposed reforms, it did examine the published ideas and offers several overarching comments to further these reform goals.

Improving Accountability

Science requires substantial societal investment, putting it in competition with other important activities that also contribute to the public good. The public therefore has the right to accountability in the use and management of resources allocated to the pursuit of scientific knowledge for the common good. The need for accountability is all the more heightened when publicly funded research also depends on the participation of human subjects. In

EXECUTIVE SUMMARY

carrying out President Obama's charge to assess the degree to which current federally funded research protects human subjects from harm or unethical treatment, the Commission encountered a significant challenge in ascertaining federal investment in human subjects research. Internal department or agency-specific systems to track human subjects research are generally available, although they vary widely in the basic information they maintain and differ considerably in the information they can readily retrieve or make available publicly or online. To accurately track and assess the volume and scope of human subjects research and to determine whether protections are in place, there must be better data and more ready availability of information.

Recommendation 1: Improve Accountability through Public Access

To enhance public access to basic information about federal government-funded human subjects research, each department or agency that supports human subjects research should make publicly available a core set of data elements for their research programs—title, investigator, location, and funding—through their own systems or a trans-agency system. The Office for Human Research Protections or another designated central organizing agency should support and administer a central web-based portal linking to each departmental or agency system. This should not preclude the prospective development of a unified federal database that may ultimately be more cost-effective and efficient.

The Commission also encourages additional research into the effectiveness of human subjects protection standards to obtain empirical data through which to assess such protections. Such data could, for example, illuminate issues pertinent to research site selection or assist in promoting the effectiveness of community engagement—both topics the Commission believes could use improvement.

Recommendation 2: Improve Accountability through Expanded Research

To evaluate the effectiveness of procedural standards embedded in current human subjects protection regulations, the federal government should support the development of systematic approaches to assess the effectiveness of human subjects protections and should expand support for research related to ethical and social consideration of human subjects protection.

Treating and Compensating for Research-Related Injury

Those who sponsor or engage in human subjects research have an ethical obligation to protect those who volunteer as research subjects. Almost all other developed nations have instituted policies to require treatment, or compensation for treatment, for injuries suffered by research subjects. The Panel advised the Commission to recommend that the United States establish a system to assure compensation for the medical care of human subjects harmed in the course of biomedical research. However, the Commission believes that before altering the current approach to compensation for injuries sustained during federally funded research, the nature and scope of harms that remain unaddressed must be assessed.

Recommendation 3: Treating and Compensating for Research-Related Injury

Because subjects harmed in the course of human research should not individually bear the costs of care required to treat harms resulting directly from that research, the federal government, through the Office of Science and Technology Policy or the Department of Health and Human Services, should move expeditiously to study the issue of research-related injuries to determine if there is a need for a national system of compensation or treatment for research-related injuries. If so, the Department of Health and Human Services, as the primary funder of biomedical research, should conduct a pilot study to evaluate possible program mechanisms.

The Commission stresses that it is important to recognize the limits of current models for providing compensation, like the National Vaccine Injury Compensation program, and also the various means by which the government may satisfy the ethical obligation to compensate individuals who suffer research-related injuries in a federally funded study. While there are systems already in place for some government research, the Commission recommends a study to evaluate future options and outlines many questions to be considered. It also recognizes that several national bodies have made similar recommendations in the past. Given the seriousness of the ethical concern, and these past efforts, the Commission encourages the government to follow up publicly with its response.

EXECUTIVE SUMMARY

Recommendation 4: Treating and Compensating for Research-Related Injury Follow Up

The Commission recognizes that previous presidentially appointed bioethics commissions and other duly appointed advisory bodies have made similar recommendations regarding compensation or treatment for research-related injuries; yet no clear response by the federal government has been issued. Therefore, the federal government, through the Office of Science and Technology Policy or the Department of Health and Human Services, should publicly release reasons for changing or maintaining the status quo.

Creating a Culture of Responsibility: Human Research Protections as Professional Standards

The Commission heard from a wide range of research professionals that the procedural requirements of human subjects regulations are often viewed as unwelcome bureaucratic obstacles to conducting research. The density of some of these requirements can obscure their justification and routinized interpretation can create distance between the underlying ethical principles and how they are viewed and implemented by institutional review boards and the research community. The Commission too recognizes that there is often a fundamental distinction between ethical principles (and the personal responsibilities that must be exercised to effect them), and procedural or policy strategies to apply and implement these principles. While tension between principles and procedures is, in some ways, perennial, the Commission believes that specific steps can be taken now to deflect the tilt that some see favoring process over principle. Two of these recommendations are directed to government specifically, and a third more generally relates to education and the duty to all engaged in the research enterprise.

Recommendation 5: Make the Ethical Underpinnings of Regulations More Explicit

To promote a better understanding of the context and rationale for applicable regulatory requirements, the Department of Health and Human Services or the Office of Science and Technology Policy should ensure that the ethical underpinnings of regulations are made explicit. This goal is also instrumental to the current effort to enhance protections while reducing burden through reform of the Common Rule and related Food and Drug Administration regulations. (See *Promoting Current Federal Reform Efforts* below.) Following the principle of regulatory parsimony, regulatory provisions should be rationalized so that fundamental, core ethical standards are clearly articulated.

Recommendation 6: Amend the Common Rule to Address Investigator Responsibilities

The Common Rule should be revised to include a section directly addressing the responsibilities of investigators. Doing so would bring it into harmony with the Food and Drug Administration regulations for clinical research and international standards that make the obligations of individual researchers more explicit, and contribute to building a stronger culture of responsibility among investigators.

Recommendation 7: Expand Ethics Discourse and Education

To ensure the ethical design and conduct of human subjects research, universities, professional societies, licensing bodies, and journals should adopt more effective ways of integrating a lively understanding of personal responsibility into professional research practice. Rigorous courses in bioethics and human subjects research at the undergraduate as well as graduate and professional levels should be developed and expanded to include ongoing engagement and case reviews for investigators at all levels of experience.

Respecting Equivalent Protections

Research supported by the federal government is subject to the same regulatory requirements domestically as well as internationally. Research collaborators and partners in other countries who are funded or supported by Common Rule agencies file an assurance that they will comply with these requirements, regardless of overlapping or more protective standards that may exist in the country where the research is conducted. At the same time, U.S. regulations governing the protection of human research subjects delineated in the Common Rule have long permitted U.S. departments and agencies supporting or conducting research to recognize and accept procedures from foreign countries that may differ from those delineated in U.S. regulations as long as they provide "protections that are at least equivalent" to those in the Common Rule. Yet U.S. departments and agencies have rarely, if ever, exercised the authority to accept any foreign country's procedures as equivalent. Instead they sometimes insist that all U.S. procedural details must be met, in all cases, regardless of the effectiveness or similarity of foreign requirements. This insistence on both the spirit and letter of U.S. regulatory constraint fails to recognize or respect that many nations today have systems to protect human subjects that are as good, or perhaps more stringent, than our own. Despite numerous efforts to clarify or resolve the meaning and interpretation of "equivalent protections," no comprehensive policy has emerged for determining when equivalent protections exist.

Recommendation 8: Respect Equivalent Protections

The federal government, through the Office for Human Research Protections, should adopt or revise the 2003 Health and Human Services Equivalent Protections Working Group's articulation of the protections afforded by the specific procedural requirements of the Common Rule. It should use these requirements to develop a process for evaluating requests from foreign governments and other non-U.S. institutions to determine if their laws, regulations, and procedures can be recognized as providing equivalent protections to research subjects.

Promoting Community Engagement

The Panel directed the Commission's attention to the value of community engagement and demonstration of respect for cultural differences that are compatible with the ethical conduct of human subjects research. These values are applicable to research conducted both domestically and abroad. Effective community engagement provides an additional layer of safeguards by providing the community with opportunities to thoroughly weigh and accept or reject the risks and benefits of research activities, discover possible implications of research that might have unintended consequences to the host community, and independently debate the effectiveness of research protections. Interactive and ongoing dialogue between communities and research teams allows for the integration of community norms, beliefs, customs, and cultural sensitivities into research activities. The guidelines enumerated in the Joint United Nations Programme on HIV/AIDS and the AVAC *Good Participatory Practice Guidelines*, for example, provide a standardized framework for implementing community engagement activities across a wide range of research. The Commission believes these, and related documents, should be evaluated and guidance provided by the government on effective community engagement strategies for all human subjects research.

Recommendation 9: Promote Community Engagement

The federal government, through the Office for Human Research Protections and authorized research funders, should support further evaluation and specification of the Joint United Nations Programme on HIV/AIDS and the AVAC *Good Participatory Practice Guidelines* with the aim of providing a standardized framework for those community engagement practices that would further the protection and ethical treatment of human subjects in all areas of research. Research should be conducted to prospectively evaluate the effectiveness of this framework and strengthen it after it is developed.

Justifying Site Selection

Careful selection of sites for research is important for two sets of reasons. First, the ethical criteria for how subjects must be treated narrows the selection of sites to those that allow for the ethical treatment of subjects. Second, as the *Belmont Report* states, "selection of research subjects needs to be scrutinized in order to determine whether some classes are being systematically selected because of their easy availability, their compromised position, or their manipulability, rather than for reasons directly related to the problem being studied." Thus, careful examination of site selection is extremely important and critical to ensuring that research is done ethically and participants are protected from harm or unethical treatment. Some domestic and international settings present challenges that increase concern about exploitation of human subjects. One proposed strategy for minimizing the potential of exploitation when research is done in low-income communities—whether domestic or international—is to ensure that the proposed study is responsive to the medical, as well as other, needs of the local community or communities. The ethical requirement of responsiveness to local communities needs to be further developed and implemented for responsiveness to become a clearly justified criterion for site selection.

Recommendation 10: Ensure Capacity to Protect Human Subjects

Funders of research should determine that researchers and the sites that they propose to select for their research have the capacity—or can achieve the capacity contemporaneously with the conduct of the research—to support protection of all human subjects.

Recommendation 11: Evaluate Responsiveness to Local Needs as a Condition for Ethical Site Selection

The federal government, through the Office for Human Research Protections and federal funding agencies, should develop and evaluate justifications and operational criteria for ethical site selection, taking into consideration the extent to which site selection can and should respond to the needs of a broader community or communities. The Office for Human Research Protections should produce, and other agencies should consider developing, guidance for investigators.

Ensuring Ethical Study Design

Study design, particularly in clinical research, is another area where concerns about exploitation have arisen in the past. The Commission reviewed issues surrounding use of placebo and other comparator arms in randomized clinical trials conducted in locations that do not have access to the highest standard of care. It found consensus around a "middle ground" to guide researchers designing clinical trials that expose subjects to interventions or conditions that may not be viewed as the best available standard of care but nonetheless provide potential for benefit to the local population. The Commission's proposed framework, rooted in the thinking that has developed in the literature over several decades, provides a pathway to ensuring that research subjects' interests are protected.

Recommendation 12: Ensure Ethical Study Design for Control Trials

When assessing how to reconcile the requirements of rigorous study design with the interests of research subjects, a nuanced approach is recommended that permits subjects to receive a placebo or an active agent that otherwise might not represent the "best-proven" approach when the site selected is ethically justifiable and the following conditions are met: a) the "best-proven" intervention is not known to be the best for a particular population due to local infrastructure, behavioral, genetic, or other relevant circumstances; and b) the scientific rationale *and* the ethical justification for the study design have undergone careful review to ensure all of the following: i) use of placebo or other comparators is of limited duration; ii) subjects are carefully monitored; iii) rescue measures are in place should serious symptoms develop; and iv) there are established withdrawal criteria in place for subjects who experience adverse events.

Promoting Current Federal Reform Efforts

The Commission commends efforts already underway to reform federal policy for the protection of human subjects and recommends that these efforts be advanced. In particular, the Commission endorses the following proposals presented in the Advance Notice of Public Rulemaking issued in July 2011 by the Office of the Secretary of the Department of Health and Human Services (HHS) in coordination with the Office of Science and Technology Policy (OSTP).

Recommendation 13: Promoting Current Federal Reform Efforts

The Commission supports the federal government's proposed reforms to:

a) Restructure research oversight to appropriately calibrate the level and intensity of the review activities with the level of risk to human subjects;

b) Eliminate continuing review for certain lower-risk studies and regularly update the list of research categories that may undergo expedited review;

c) Reduce unnecessary, duplicative, or redundant institutional review board review in multi-site studies. Regardless of the process used to review and approve studies, institutions should retain responsibility for ensuring that human subjects are protected at their location as protection of human subjects includes much more than institutional review board review. The use of a single institutional review board of record should be made the regulatory default unless institutions or investigators have sufficient justification to act otherwise;

d) Make available standardized consent form templates with clear language understandable to subjects;

e) Harmonize the Common Rule and existing regulations of the Food and Drug Administration, and require that all federal agencies conducting human subjects research adopt human subjects regulations that are consistent with the ethical requirements of the Common Rule; and

f) Work toward developing an interoperable or compatible data collection system for adverse event reporting across the federal government.

Following Up

The Commission recognizes that several of these recommendations have been made by presidentially appointed bioethics commissions and other duly appointed government advisory bodies in the past, and it found no clear response by the federal government to many of them. For example, a number of commissions, including the National Commission for the Protection of Human Subjects of Biomedical and Behavioral Research, the National Bioethics Advisory Commission, the Presidential Commission for the Study of Ethical Problems in Medicine and Biomedical and Behavioral Research, and the President's Advisory Committee on Human Radiation Experiments, made recommendations endorsing compensation for subjects for injuries arising from research.[7] Both National Bioethics Advisory Commission and Advisory Committee on Human Radiation Experiments recommended recognition of equivalent protections.[8] In addition, National Bioethics Advisory Commission addressed many of the same issues raised in the recommendations in this report, such as community engagement, ethics training, and the importance of an expanded research agenda addressing human research-related issues.[9]

Recommendation 14: Responding to Recommendations

The Commission recommends that the Office of Science and Technology Policy or another appropriate entity or entities within the government respond with changes to the status quo or, if no changes are proposed, reasons for maintaining the status quo with regard to the recommendations below. Possible departments or agencies to lead the efforts include the Department of Health and Human Services, the Office for Human Research Protections, and the National Institutes of Health, as well as other funders and regulators.

Table ES.1 Recommendation Follow-up Summary

RECOMMENDATION NUMBER[†]	SUMMARY	OFFICE
1	Increase accountability through online access to basic human subjects research data.	OHRP/all departments and agencies that support human subjects research
2	Support the development of systematic approaches to assess the effectiveness of human subjects protections and expand support for research related to ethical and social consideration of human subjects protection.	OHRP/all departments and agencies that support human subjects research
3	Study research-related injuries to determine if there is a need for a national system of compensation or treatment for research-related injuries because subjects harmed in the course of human research should not individually bear the costs of care required to treat harms resulting directly from that research.	OSTP/HHS
4	Publicly release reasons for changing or maintaining the status quo regarding compensation or treatment for research-related injuries.	OSTP/HHS
5	Explicate the ethical underpinnings for human subjects protection requirements.	HHS/OSTP
6	Add responsibilities of investigators to the Common Rule.	HHS/OSTP
8	Adopt or revise the 2003 Department of Health and Human Services Equivalent Protections Working Group's analysis and develop a process for evaluating requests from foreign governments and other non-U.S. institutions for determinations of equivalent protections.	OHRP
9	Support further evaluation of the UNAIDS/AVAC Guidelines to provide a standardized framework for community engagement practices across research fields.	OHRP
11	Support research to develop and evaluate justifications and operational criteria for ethical site selection.	OHRP/all departments and agencies that support human subjects research
13	Develop proposed regulations to reform the current Common Rule.	OSTP/OHRP
14	Follow up.	OSTP/other appropriate entity

[†] Listed here are recommendations directed to the federal government only.

A response need not be unduly lengthy or be provided by a single department, agency, or division, but the public should know whether the federal government intends to move forward, and if so in what way, with any or all of these recommendations. (See Table ES.1 for a summary of recommendations directed towards the federal government.)

CHAPTER 1
Introduction

Research is not only important as a means of advancing knowledge; it is also a core component of America's growth and prosperity in human health, energy, defense, education, and countless other components of daily life. Social progress depends on new discoveries and new ways of thinking about old problems. Yet, as the philosopher Hans Jonas observed, "society would indeed be threatened by the erosion of those moral values whose loss, possibly caused by too ruthless a pursuit of scientific progress, would make its most dazzling triumphs not worth having."[10]

The relationship between scientific progress and morality is not a new concern in the United States. Writing nearly 200 years before Jonas, and in the midst of the hard work of building the new nation, Benjamin Franklin reflected that "[t]he rapid progress *true* science now makes, occasions my regretting sometimes that I was born so soon." He continued:

> "It is impossible to imagine the height to which may be carried, in a thousand years, the power of man over matter . . . . Agriculture may diminish its labor and double its produce; all diseases may by sure means be prevented or cured, not excepting even that of old age, and our lives lengthened at pleasure even beyond the antediluvian standard. O that moral science were in as fair a way of improvement, that men would cease to be wolves to one another, and that human beings would at length learn what they now improperly call humanity!"[11]

Surely Franklin would not be wholly satisfied in how far our society or the world has progressed in conquering "wolfish humanity," and neither should society. Yet Franklin and the other American founders believed that the values of individual freedom of speech, conscience, and inquiry—along with a dedication to pursuing the common good—held great promise for a republic dedicated to progress in both science and morality, and that science and morality inform one another by challenging dogmatism in either realm. What Franklin called "moral science" is what we would today call "ethics."[12] Franklin clearly intended that "moral science" (ethics) should inform empirical science. Thus, using a contemporary play on Franklin's phrase, one can say that the challenge of "moral science" (i.e., pursuing science in a morally justified manner) is one that every generation must take up again.[13]

INTRODUCTION

There is no more acute instance of this challenge than research involving human subjects, upon which knowledge and discovery often depend. Human research subjects, in most cases, must be informed volunteers who are willing to allow their bodies or personal information to be used by researchers to craft new hypotheses, infer plausible explanations and predictions, and test theories.[14] Pursuit of these goals sometimes offers no direct prospect of benefit to the human research subject. Early stage translational research serves to test physiological effects or biological functions of new drugs and medical interventions but is not necessarily designed to benefit subjects. Sometimes research subjects experience serious adverse health effects as a result of participation in trials.[15] Research in other fields, including housing, social work, and criminology, typically poses fewer physical risks, but may pose substantial social, psychological, and economic risks for human subjects. Such research also aims to improve the lives of later generations without offering any direct or measurable benefit to those who participate as research subjects.

Regardless of whether research offers the prospect of direct benefit to human subjects, long-standing ethical principles constrain the unfettered pursuit of knowledge. For research with human subjects to be ethical, volunteers must be treated fairly and with respect, subjected only to reasonable risks from which proportionate humanitarian benefit can be obtained, and not treated as mere means to the ends of others. Experimentation is rooted in uncertainty, and human subjects cannot be immunized from all physical risks. Nonetheless, these stated boundaries dictate that anticipated and avoidable harms must be eliminated, informed consent must be obtained in most cases, and the burdens and benefits of research must be equitably shared.

From time to time, society has been reminded of the need to revisit the rules applied to research with human subjects. The need for reassessment might arise from challenges emerging from novel scientific advances or from a perceived mismatch between ethical principles and how they are implemented in practice. Revelations of abuses also have been a driving force in reconsideration of the policies and procedures for protecting the subjects of research.

In the past few years, several factors have converged to compel a contemporary review of human subjects protection policies and practices. The Presidential Commission for the Study of Bioethical Issues (the Commission) initiated this review in response to a charge from President Obama to undertake "a thorough review of [current] human subjects protection to determine if Federal regulations and international standards adequately guard the health and well-being of participants in scientific studies supported by the Federal Government."[16]

This request came paired with the President's charge to conduct an investigation of new revelations about medical research supported by the United States and conducted in Guatemala between 1946 and 1948. Some of that research involved the deliberate exposure of people to sexually transmitted diseases (STDs) without their consent. Subjects, including soldiers, prisoners, psychiatric patients in a state-run institution, and commercial sex workers, were exposed to syphilis, gonorrhea, and chancroid. Serology diagnostic testing involved the previous four groups as well as children, U.S. Servicemen stationed in Guatemala, and leprosarium patients. In October 2010, President Obama expressed "deep regret" to the President of Guatemala for this research, and affirmed the federal government's "unwavering commitment to ensure that all human medical studies conducted today meet exacting" standards for the protection of human subjects.[17]

The results of the Commission's investigation into the Guatemala experiments were released in September 2011 in its report, *"Ethically Impossible" STD Research in Guatemala from 1946 to 1948*, the findings of which are summarized below.[18] That report provided a historical account and ethical assessment of the Guatemala experiments. It uncovered and contextualized as much as could be known about experiments that took place nearly 65 years ago. It also aimed to inform current and continuing efforts to protect the rights and welfare of the subjects of U.S.-sponsored or -conducted research. In *"Ethically Impossible,"* the Commission recognized that U.S. and international policies and practices governing human subjects research have evolved in the time since the Guatemala experiments from an informal set of principles, based primarily on medical ethics and the doctor-patient relationship, to a highly structured oversight system codified in regulation and statute.

However, the uncovering of the 1940s STD studies in Guatemala prompted President Obama to ask whether such unethical research could be conducted in today's research environment using U.S. federal funds. When the President asked the Commission to undertake this review of human subjects protection

INTRODUCTION | 1

standards, he recognized, as do we all, the sacred trust and responsibility that society has to ensure that human research subjects are protected from harm and unethical treatment.[19]

The Need to Assess the Contemporary Environment

Since the 1940s, the research enterprise and the system of protection for human subjects have continuously evolved. Abuses uncovered in the 1960s and 1970s led to the development of the current U.S. regulations for the protection of human subjects. These regulations combine with longstanding professional norms and obligations to map the ethical boundaries for research today. For federally funded research, the rules have remained largely unchanged since at least 1991. For some agencies, the regulations are even older. For example, explicit Department of Health and Human Services (HHS; then the Department of Health, Education, and Welfare) policy requirements began in 1953 with the opening of the National Institutes of Health (NIH) Clinical Center (the agency's internal research hospital).[20] They expanded with statutory requirements for informed consent in Food and Drug Administration (FDA) regulated clinical trials and other HHS policy requirements in the 1960s, and were further modified with the regulations for government-sponsored medical research promulgated in 1972.[21]

> Regarding the changing geography of clinical research, "[w]e are seeing a massive shift in the conduct of research.... [W]hat we are seeing is a large shift in economics and finance occurring at a setting in which the marketing of medical products has become global in every respect."
>
> Dr. Robert M. Califf, Vice Chancellor for Clinical Research, Duke University Professor of Medicine, Duke University Medical Center, Director, Duke Translational Medicine Institute, speaking to the Commission on March 1, 2011.

The decades that have elapsed since the first national bioethics commission began an evaluation of U.S. human subjects protections in 1974 have seen dramatic changes in the research enterprise and further development of our system for protecting human subjects.[22] Research beyond public health and medicine, in social science and related fields, can involve thousands of research subjects through increasingly accessible survey tools and methodologies that expand experimental rigor.[23] Much like the globalization of business and other sectors of our economy and culture, research with human subjects is now a global enterprise (see Figure 1.1).

21

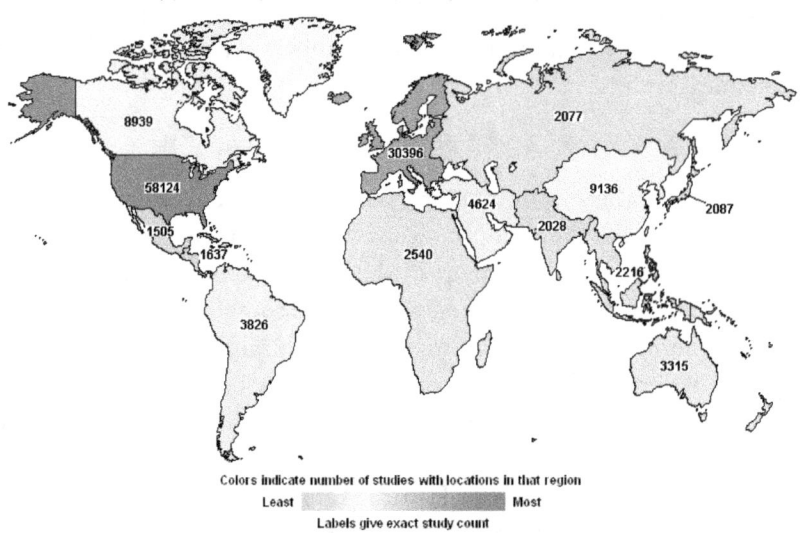

Figure 1.1: National Institutes of Health. Map of All Studies in ClinicalTrials.gov. (n.d.). Retrieved from http://www.clinicaltrials.gov/ct2/search/map/click?map.x=322&map.y=305 (accessed December 5, 2011).

In the health sector, large-scale, multi-site, and multi-national clinical trials and management by contract research organizations (third parties to whom research sponsors outsource many of the administrative needs for research) are just two markers of this increasingly decentralized enterprise. Biomedical research is rapidly and inexorably expanding internationally.[24] It increasingly addresses diseases more prevalent in other, sometimes developing, nations. Treatment naïve subjects (those not yet exposed to myriad pharmaceutical interventions), larger study populations, and the need to perform research in certain countries for both marketing and scientific purposes, further explain decisions to locate research in international settings.[25] With the exception of the goals of product marketing, these factors are relevant to both publicly and privately funded research activities.

> "Looking at the change in global clinical trials, I think many people have rightly said that pendulum is swinging from where the majority of clinical trials were done in North America and Western Europe, to the point now where they are being done all over, to the point where they are going to be done in other parts of the world...."
>
> Dr. Murray M. Lumpkin, Deputy Commissioner for International Programs, FDA, speaking to the Commission on May 18, 2011.

INTRODUCTION I

Research Across Borders

Recognizing the increasing involvement of foreign sites and partners in human subjects research, the Commission organized the International Research Panel (Panel), consisting of experts in bioethics and biomedical research hailing from 10 countries: Argentina, Belgium, Brazil, China, Egypt, Guatemala, India, Russia, Uganda, and the United States.[26] The Panel, chaired by Commission Chair Amy Gutmann, met on three day-long occasions to discuss research standards and practices around the globe. In their discussions, Panel members drew upon their individual expertise and decades of experience conducting research and developing policy to protect human subjects.

The Panel presented its findings and recommendations, which focused on biomedical research, to the Commission in its report, *Research Across Borders: Proceedings of the International Research Panel of the Presidential Commission for the Study of Bioethical Issues.*[27] The Panel reached consensus on several issues. The United States and many other nations, the members agreed, have made significant progress in developing measures to protect human subjects in research over the last 50 years. Still, the rules, standards, and practices inside and outside the United States continue to vary greatly.[28] The Commission's research on international standards confirmed these findings.

The Panel concluded that researchers must demonstrate respect for human subjects and their communities in all phases of study design and implementation, and wherever research occurs. Recognizing different cultural standards and practices through community engagement that are compatible with the ethical treatment of human subjects is an important way of demonstrating respect across national boundaries, just as it is an important way of demonstrating respect within a diverse society. The Panel also stressed the importance of ongoing international dialogue between the United States and international bodies. As many within the United States and international partners have said before, the Panel agreed that U.S. and foreign investigators would benefit from clarification of the U.S. regulatory exception for foreign "protections that are at least equivalent to those" in the United States ("equivalent protections" found at 45 C.F.R. § 46.101(h)), and how it can be applied.

> **INTERNATIONAL COLLABORATION: RV144 TRIAL**
>
> Working with the Thai Ministry of Public Health (MOPH), U.S. and Thai researchers tested an HIV vaccine (RV144) that showed a 31.2% efficacy rate in preventing HIV transmission in a study sponsored by the U.S. Army and primarily funded by NIH.
>
> The Panel considered this case study in its proceedings and focused its attention on the collaboration between the two countries. Of note, the vaccine developers guaranteed from the outset that, if the vaccine proved effective, Thailand would get it at a discount, as well as rights to manufacture it locally. In addition, naming a Thai MOPH official as one of the principal investigators in the study assured the Thai government that its interests would be honored.
>
> Source: Presidential Commission for the Study of Bioethical Issues. (September 2011). Research Across Borders: Proceedings of the International Research Panel of the Presidential Commission for the Study of Bioethical Issues, pp. 50-53. Washington, DC: Presidential Commission for the Study of Bioethical Issues; AVAC (2010). AVAC Report 2010: Turning the Page, p. 25. AVAC. Retrieved from http://www.avac.org/ht/a/GetDocumentAction/i/28305.

The Panel proposed increased accountability by expanding access to public information about research.[29] "Greater efforts are needed," the Panel concluded, "to enhance transparency, monitor ongoing research, and hold researchers and institutions responsible and accountable for violations of applicable rules, standards, and practices." Specifically, the Panel suggested that the government require all greater than minimal risk research to be registered and results reported in a system such as ClinicalTrials.gov (a federally sponsored on-line registry of most clinical trials conducted in or outside of the United States).[30]

The Panel also highlighted one of the widest disparities between the United States and other nations—many other countries provide greater protection for human subjects by ensuring treatment for research-related injuries or compensation for the costs of research-related treatment. The Panel noted that most other developed countries require sponsors, investigators, or others engaged in research to provide such treatment or reimbursement free of charge to the subject. The Commission's research reached similar conclusions (see Appendix IV). In light of these facts, and the strong ethical case for such protection, the Panel recommended that the federal government explore whether revision of its rules is needed to ensure compensation for the medical costs of research-related injuries.[31]

INTRODUCTION

Finally, the Panel echoed a view voiced by many of the guest speakers and individuals who provided written comments to the Commission. Reflecting what might be called "process fatigue," the group endorsed ongoing efforts to harmonize and clarify existing rules to better protect human subjects over creating new rules.[32]

The Panel's findings and recommendations gave the Commission important insight into the practice of human subjects research internationally and appreciation of the capacity of the current system to protect human subjects from harm and unethical treatment domestically as well as internationally. Furthermore, the Panel reinforced a conclusion the Commission reached in its own research—namely, that the ethical principles constraining pursuit of knowledge with human subjects apply regardless of where research takes place or who funds it.[33] Procedural differences can, and sometimes should, affect the implementation of these principles, but the fundamental ethical foundations and boundaries of ethically sound research apply to all domestic, international, public, and private research.

Contemporaneous Reviews of the Human Subjects Protection System

Shortly after the Commission began its work in this area, the Office of the Secretary of HHS, in coordination with the White House Office of Science and Technology Policy (OSTP) issued an Advance Notice of Proposed Rulemaking, signaling the intent to consider comments on how to modernize the U.S. regulations governing human subjects research to make them more effective. This ANPRM, issued July 26, 2011, notes that the current HHS regulations[34] also signed onto by 17 other federal departments and agencies (i.e. the Common Rule), were developed more than 20 years ago, "when research was predominantly conducted at universities, colleges, and medical institutions, and each study generally took place at only a single site."[35]

In addition to changes over time in who is conducting research and where, the ANPRM also notes other trends in an evolving research enterprise, such as the expansion of health services research, research in the social and behavioral sciences, and research involving databases, the Internet, biological specimens, and genomics. The ANPRM sought comments on "how to better protect human subjects who are involved in research, while facilitating valuable research and reducing burden, delay, and ambiguity for investigators."[36]

Thus, in undertaking its review in response to the President's charge, the Commission also considered aspects of the ANPRM for which it could make an informed contribution and reviewed ongoing efforts to modernize and clarify existing rules.

About this Report

President Obama gave the Commission nine months to thoroughly investigate the activities in Guatemala and, following the appointment of the Panel in March 2011, nine months to complete this contemporary review. The Commission held four multi-day, public meetings to address the President's requests. Meeting speakers addressed a range of U.S. and international policies, rules, regulations, and publicly enunciated principles, providing perspectives from a wide array of professional and institutional organizations. The Commission also participated in deliberative discussion with members and the audience. It received more than 300 comments on this work.[37]

In the course of its review, the Commission studied human subjects research protections around the globe. A brief summary of some of those standards is made readily available through the website of the Office for Human Research Protections. The Commission also asked federal departments and agencies for information about their policies and practices to protect human research subjects.[38] Department and agency liaisons provided extensive information concerning relevant policies used during the planning, execution, and oversight of research. They also provided valuable practical insight into how each department or agency supports and conducts research, as well as the efforts they take to ensure that human subjects are protected.

The Commission's deliberations took into account the work of numerous preceding bodies. The 1995 *Final Report of the Advisory Committee on Human Radiation Experiments* offered 18 detailed and thoughtful recommendations regarding past abuses and remedies for participants injured in government-sponsored scientific studies as well as empirically based ways to improve protection for human subjects in current and future studies. These included specific recommendations for ongoing debate through public fora.[39] Many of these recommendations have since been implemented, and all of them raised important considerations to guide the Commission in its work.

INTRODUCTION

In 2001, the National Bioethics Advisory Commission issued a comprehensive report on the existing system to protect human research subjects that included more than 30 specific recommendations; it also issued a report on research in international settings with 28 recommendations.[40] In 2003, the Institute of Medicine (IOM) issued *Responsible Research: A Systems Approach to Protecting Research Participants*, with recommendations to extend the oversight system to all research, regardless of funding source or research setting.[41] Additional IOM recommendations focused on the duties and functions of Institutional Review Boards, conflict of interest rules, the informed consent process, compensation for research-related injuries, levels of ethics review, and the need for quality improvement. Foreshadowing a perspective that the Commission heard repeatedly in this investigation, and voiced previously in the context of synthetic biology, IOM Chair Daniel Federman wrote, "[w]e do not, however, urge a permanent accretion of new regulations and bureaucracy."[42] Rather, Federman urged periodic reexamination of the problems and challenges facing the oversight system to better appreciate its appropriateness in "minimiz[ing] harm while enabling the benefits of progress to emerge."[43]

Also during the last decade, the Secretary's Advisory Committee on Human Research Protections (SACHRP), which advises the Secretary of HHS, has provided extensive guidance to the government on human subjects research protections and practices, including but not limited to recommendations regarding current rules for research involving children, prisoners, and other vulnerable populations; informed consent practices; community engagement; regulatory burden; and the need for harmonization of human subjects regulations.[44] Collectively, all of this work assisted the Commission.

In Chapter 2, the Commission provides its response to the question of whether current protections are adequate to protect subjects from harm or unethical treatment. Chapter 3 includes further findings and recommendations. Additional background material appears in the appendices.

CHAPTER 2
Assessing the Current System

President Obama charged the Commission to determine whether "current rules for research participants protect people from harm or unethical treatment, domestically as well as internationally" and to undertake "a thorough review of human subjects protection to determine if federal regulations and international standards adequately guard the health and well-being of participants in scientific studies supported by the federal government."[45] In making this request the President sought assurance that the rules governing federal research today adequately guard against the abuses perpetrated by the U.S. Public Health Service in Guatemala in the 1940s. The President also asked for assurance that current rules protect people from harm or unethical treatment, no matter where in the world U.S.-supported research occurs.

In its previous report, *"Ethically Impossible" STD Research in Guatemala from 1946 to 1948*, the Commission detailed how researchers in the 1940s failed to protect subjects from harm and unethical treatment in research supported by the Public Health Service and the government of Guatemala.[46] In that case, investigators enrolled people in studies that involved intentional exposure to STDs without informing them of risks or seeking their consent. The study design, and the investigators' actions in executing it, relied on flawed methodology and failed to minimize risks or maximize benefits for the majority of subjects and the community. Investigators intentionally hid information from the subjects, the public, and others who might have questioned their methods or aims.

In 1946, when the research in Guatemala began, no federal laws, regulations, guidelines, or explicit international standards protected human subjects from these abuses. However, professional standards for physicians that had been promulgated by the American Medical Association, and the researchers' own conduct in a previous U.S.-based STD trial that they had conducted in 1943 and 1944, demonstrated their awareness that participants should not be exposed to undue risk and should be informed volunteers. Given these facts, the Commission concluded that many of the investigators and others in the chain of command supporting the research failed to protect subjects from harm and unethical treatment and did not abide by ethical constraints that they knew applied to their work.

ASSESSING THE CURRENT SYSTEM   II

Since the 1940s, however, major changes in the oversight and practice of research with human subjects have created a vastly different world for researchers and subjects alike. Federal laws, regulations, and guidelines; numerous transnational standards; and many similar laws and guidelines in other countries, enumerate and impose specific provisions to protect human subjects. Today's federal rules reflect the now widely and long-recognized view that researchers must demonstrate respect for all human research subjects, including minimizing the risks and maximizing the benefits of research before it begins and respecting each person's right to give his or her informed consent or its moral equivalent. These norms also are reflected in the terms of international human rights treaties, such as the International Covenant on Civil and Political Rights, to which the United States is a signatory.[47] Americans, like many people and nations, hold these norms to be fundamental moral duties owed to every person, each of whom is entitled to respect by virtue of their unique status as moral agents.

Federal statutes and regulations require informed consent from volunteers, independent ethical review, equitable subject selection, confirmation of scientific validity, and minimization of risks to subjects, among other essential limits.[48] The Common Rule extends these requirements to most federally funded scientific studies regardless of where they occur.[49] In addition, FDA regulations for protecting human subjects apply to all clinical research on drugs, including biologics, and devices, regardless of the source of funding (i.e. public or private sources).[50] (See Figure 2.1.) The history of

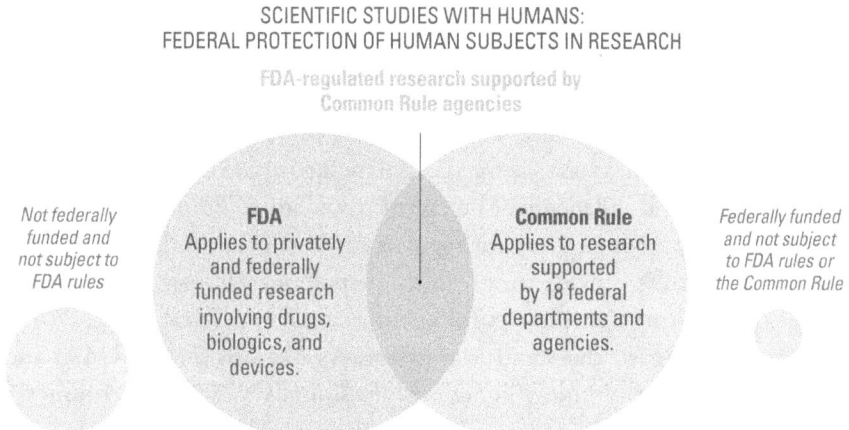

Figure 2.1: Federal Protection of Human Subjects in Research

these rules, which have evolved over several decades, is well documented. Together, they extend to most, although not all, research conducted in the United States and nearly all research funded with public monies outside of the country.[51]

Regulatory standards are not the sole or sufficient means to ensure that human subjects are protected from harm or unethical treatment. There are long and distinguished philosophical traditions, traceable in Western traditions, for example, to Aristotle's moral psychology and to a Kantian ethics of acting out of moral duty, that emphasize the need to cultivate individual moral character.[52] For research, this focus on virtuous character translates into a focus on the internal ethical motivation of individual investigators, not only the rules and regulations that externally motivate investigators toward compliance.

Although there are disagreements about how much reliance should be placed on the virtuous researcher, and although there are notorious instances in which virtue alone has not sufficed as a bulwark of research protections, it is undeniable that any compliance regimen is more effective and efficient when researchers have internalized a strong sense of personal responsibility as part of their professional calling. A strong sense of personal responsibility that supports research ethics may emerge from the individual's own moral sensibility, but it also can and should be cultivated through an education that effectively emphasizes the importance of ethics and a keen sense of social responsibility in professional life.

Professional standards for research today, as the Commission details in Chapter 3, are available but limited. Existing guidelines are found in the codes of conduct of numerous professional societies, universities, and pharmaceutical companies.[53] Embedded in these codes are the principles of respect for persons, beneficence, and justice, as demonstrated by the inclusion of safeguards such as standards for risk balancing, disclosure of investigator funding source, and independent review of payments to human subjects and investigators. Provisions calling for registration of clinical trials and publication of study results provide for greater transparency. Review requirements also seek to ensure scientific rigor and subject safety in trial design.[54] The Pharmaceutical Researchers and Manufacturers Association (PhRMA) and many of its affiliates promote general professional standards of conduct.[55]

ASSESSING THE CURRENT SYSTEM

AstraZeneca is one organization that explicitly characterizes the conduct of research ethics as a professional standard.[56] Various federal agencies provide protections in addition to those in the Common Rule or FDA rules. These include conflict of interest regulations to guide investigators in disclosure practices and privacy rules that ensure researchers comply with confidentiality requirements of the Health Insurance Portability and Accountability Act (HIPAA) Privacy Rule.[57]

With well-recognized rules (including regulations and professional standards) for consent, prospective review, scientific validity, and minimization of risks now in place, there is a starting point to answer the question of whether research subjects are adequately protected from harm and unethical treatment. However, having adequate rules in place does not, in and of itself, ensure that those rules are well implemented or, that they therefore provide adequate protection for human subjects. A far more extensive analysis than is possible now would assess whether these rules are sufficiently understood and implemented given the scope and volume of human subjects research supported or conducted by the federal government today.

The Scope and Volume of Federal Human Subjects Research

Faced with the need to assess all scientific studies with humans supported by the government, the Commission found that the scope and volume of federally supported scientific studies involving human subjects is broad and not easily identified. Available information about U.S.-funded research is sporadic, with no single listing of human subjects research available inside or outside of the government. Public information on research in medicine and health is generally more accessible than information on research in other sectors, such as education, justice, or engineering. But, overall, systemic information is very limited.

Given these limitations, the Commission sought to collect basic, project-level data about human subjects research, including study title, number and location of sites, number of subjects, and funding information directly from each Common Rule department and agency.[58] These data were compiled into its Research Project Database, and analyzed as part of its Human Subjects Research Landscape Project (see further discussion in Chapter 3 and Appendices I and II).

The Commission's data collection efforts show that federally supported human subjects research is occurring around the globe and with an enormous investment by the federal government.[59] Over 55,000 human subjects research projects (awards and individual studies) were supported in Fiscal Year 2010, mostly in medical and health-related research, but also in other fields such as education and engineering. The Commission's direct request for information from Common Rule departments and agencies revealed that some lack the means to readily identify what human subjects research they support. For example, some lack a straightforward way, such as the ability to sort relevant records in internal systems, to identify human subjects research projects. Short of reaching down into the operating components of departments and agencies and to individual program officers, which several agencies did to support the Commission's efforts, many agencies have no means to identify even basic information about agency-supported human subjects research.[60] With over six months to gather and submit project data, some departments and agencies were unable to respond fully to the Commission's request.

These systemic problems notwithstanding, the Commission found that the federal government supported approximately 55,386 human subjects research projects in Fiscal Year 2010. HHS supported almost half of all projects, approximately 26,651, with NIH supporting the largest percentage of HHS research by far, approximately 23,891 projects, or 89.6 percent of HHS studies and 43 percent of all federal studies. (See Figures 2.2, 2.3, and Tables 2.1 and I.4.) Data from earlier years, though less complete, show that the total number of human subjects projects is rising but the distribution among departments and agencies is fairly steady. HHS is the largest supporter of research with human subjects and, within it, NIH is the primary funder. (See Tables 2.1, I.4, I.9, and I.10.)

ASSESSING THE CURRENT SYSTEM

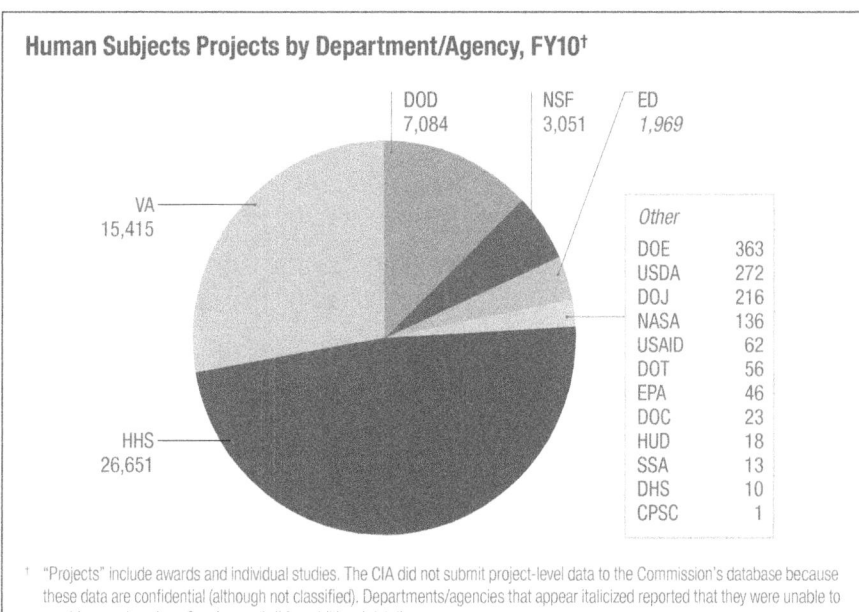

Figure 2.2

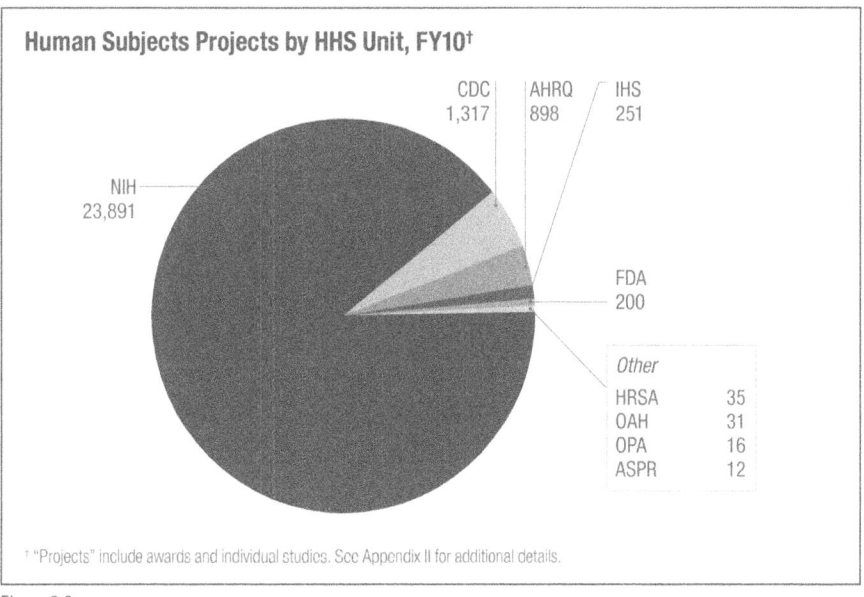

Figure 2.3

Table 2.1 Human Subjects Projects Over Time†

DEPARTMENT/ AGENCY	FY06	FY07	FY08	FY09	FY10	MEAN
HHS	25,275	25,700	25,168	26,512	26,651	25,861
VA	16,763	16,731	16,706	16,383	15,415	16,400
DOD	6,518	6,557	6,886	6,279	7,084	6,665
NSF	1,820	2,271	2,627	2,988	3,051	2,551
ED	*222*	*1,145*	*1,199*	*1,296*	*1,969*	*1,166*
DOE	407	368	371	348	363	371
USDA	118	122	220	201	272	187
DOJ	120	118	86	140	216	136
NASA	83	104	121	133	136	115
USAID	159	157	106	64	62	110
DOT	*26*	*26*	*35*	*47*	*56*	*38*
EPA	36	41	43	21	46	37
DOC	21	26	34	27	23	26
DHS‡	N/R	22	30	24	10	22
SSA	15	18	17	14	13	15
HUD	7	11	7	18	18	12
CPSC	1	1	2	1	1	1
CIA§	N/R	N/R	N/R	N/R	N/R	N/R
Total	51,591	53,418	53,658	54,496	55,386	

† "Projects" include awards and individual studies. "N/R" means that the data were not reported to the Commission. Departments/agencies that appear italicized reported that they were unable to provide complete data. See Appendix II for additional details.
‡ DHS reported that there "are no earlier data" than FY07.
§ The CIA did not submit project-level data to the Commission's database because these data are confidential (although not classified).

It is no surprise that HHS, and NIH within it, supports the largest volume of human subjects research. Research with humans is critical to advancing many fields and chief among these is medicine and health. NIH has led development of human subjects protection standards for more than 50 years since the opening of its Clinical Center in 1953.[61]

ASSESSING THE CURRENT SYSTEM

Less well known is information about the volume of research supported outside of HHS. The top five supporters of research with human subjects in Fiscal Year 2010, based on projects reported, were HHS, followed by the Department of Veterans Affairs (15,415), Department of Defense (7,084), National Science Foundation (3,051), and Department of Education (1,969). (See Table 2.1.) Research funded in these departments and agencies includes some medical/health-related research, but also largely includes research in other fields, like education and engineering.

Substantial human subjects research programs are also in place at the Department of Justice, the Department of Agriculture, and the Department of Energy. Here again, some of the research includes health-related activities, for example, measuring health effects of new technologies, but much relates to program and policy evaluation.[62] (See Table 2.2.) Although it is difficult to precisely quantify the proportion of health-related research, by relying on the mission of the departments and agencies—an imperfect but preliminary measure—approximately 42,066 projects were medical or health-related, and 13,320 projects were in other fields, including education and engineering.

> **SOCIAL, BEHAVIORAL, AND ECONOMIC SCIENCE RESEARCH METHODS**
>
> *Laboratory Experiments* – measure effects of social or physical manipulation in a controlled setting
>
> *Field Experiments* – measure effects of social or physical manipulation in a real world setting
>
> *Observations of Natural Behavior* – document descriptions of real world behavior
>
> *Interviews* – 1) Unstructured: gather open ended, qualitative data; 2) Semistructured: gather qualitative data on a specific topic; 3) Structured: gather quantitative data on a specific topic
>
> *Secondary Data Analyses* – analyze existing data for another research purpose
>
> Source: Coleman, C. et al. (2005) *The Ethics and Regulation of Research with Human Subjects.* Newark, NJ: LexisNexis.

Table 2.2 Examples of Non-Clinical Human Subjects Research

Department of Housing and Urban Development (HUD)	Observational Studies Illuminate Environmental Effects on Health
	A HUD-supported observational study examined whether a neighborhood's environment influences the health of its inhabitants. Researchers randomly assigned some participants living in public housing to receive vouchers to move to lower poverty areas, and found that moving to an area with lower poverty has a modest, but potentially significant, impact on the incidence of obesity and diabetes in residents.
Department of Energy (DOE)	International Research Improves Radiation Protection Standards
	Through the Radiation Effects Research Foundation (RERF), which is co-sponsored by the U.S. and Japanese governments, the DOE studies the long-term health impacts resulting from radiation exposure from the atomic bombs detonated in Hiroshima and Nagasaki in 1945. The RERF research has yielded improved radiation protection standards that are now employed worldwide.
Department of Justice (DOJ)	New Technologies Improve Safety By Preventing Injury
	DOJ supported a study to monitor the use of conducted energy devices or CEDs (e.g., Taser stun guns). Comparing safety outcomes of law enforcement agencies that deployed CEDs to those of agencies that did not, researchers found that the former had improved safety outcomes compared to the latter.

Sources: Ludwig, J., et al., (2011). Neighborhoods, obesity, and diabetes – A randomized social experiment. *The New England Journal of Medicine* 365(16),1509-1519; Radiation Effects Research Foundation: A Cooperative Japan-US Research Organization. (2007). [Foundation website]. Retrieved from http://www.rerf.or.jp/index_e.html; Police Executive Research Forum. (September 2009). *Comparing safety outcomes in police use-of-force cases for law enforcement agencies that have deployed Conducted Energy Devices and a matched comparison group that have not: A quasi-experimental evaluation.* Report submitted to the National Institute of Justice. Washington, D.C.: Police Executive Research Forum. Retrieved from http://www.policeforum.org/library/use-of-force/CED%20outcomes.pdf.

The Commission found, for some departments and agencies, that it was difficult to obtain comprehensive, country-specific data on where research is occurring. Some agencies limit support to domestic research and some lack authority to operate internationally. For example, the Department of Transportation and HHS's Indian Health Service, with 56 and 251 projects respectively reported for Fiscal Year 2010, limited their support to studies in the United States. (See Tables I.7 and I.8.) For these funders, all research occurred in the United States. By contrast, some departments and agencies did not identify specific countries in which research occurs. For example, readily available records were sometimes limited to saying whether work is "foreign" or "domestic," and some departments or agencies did not report where work is occurring with certainty. (See Figure 2.4. and Tables I.7 and I.8.) However, for those that did report country-specific information, the Commission found that U.S.-government research was occurring in at least 117 countries around the globe in Fiscal Year 2010 and on each inhabited

ASSESSING THE CURRENT SYSTEM   II

continent. About 4.6 percent of the 55,386 Fiscal Year 2010 government-supported projects included at least one international component (site or data collection); 65.1 percent were entirely domestic; and 30.3 percent did not report a site country. (See Table I.7.) In HHS, approximately 7.7 percent of the 26,651 Fiscal Year 2010 projects included at least one international component (site or data collection); 88.3 percent were entirely domestic. (See Table I.8.)

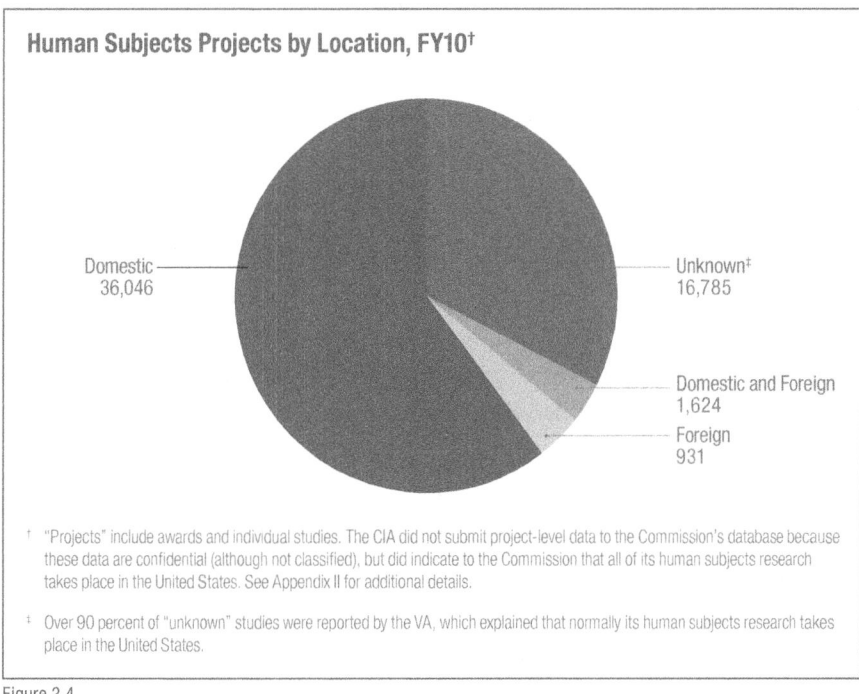

Figure 2.4

When considering responsibility for research, an important distinction exists between the government's support for intramural human subjects research (i.e., conducted directly by governmental personnel and at governmental facilities) and extramural human subjects research (i.e., conducted by academic investigators or others supported through agency grant, contract, or other mechanism). (See Tables I.2, I.3, I.5, and I.6.) Generally, departments and agencies have a far more direct line of authority and responsibility for the conduct of intramural research. It is, in a sense, "their" work. Extramural research, by contrast, is conducted by individuals outside of the government, although they are funded by the government and, for contract-funded research, operate at the behest and with specific guidance from the government. The proportion of intramural and extramural projects across the Common Rule departments and agencies varies greatly. But, generally, based on total number of projects, the government's extramural research portfolio, comprising around 54 percent of the government's total research efforts, is more extensive than its intramural portfolio. In Fiscal Year 2010, the federal government directed extramural awards to approximately 3,100 institutions, about 200 of which are located outside the United States. (See Table I.13.) Funding to these approximately 3,100 institutions totaled about $16.7 billion. (See Table I.9.)

Current Rules

Knowing the scope and volume of research is a prerequisite for assessing whether the current system is adequate for protecting human subjects from harm or unethical treatment, but another critical determinant is whether the investigators and researchers conducting this research understand and abide by the ethical rules and constraints applicable to their work.

The Common Rule agencies have various civil methods to enforce their regulations. (See Table 2.3.) Within HHS, the Office for Human Research Protections (OHRP) evaluates complaints and conducts compliance oversight site visits and not-for-cause surveillance evaluations of institutions. OHRP does not conduct investigations of noncompliance regarding research solely conducted or supported by agencies other than HHS. Those other agencies are responsible for directly investigating allegations of noncompliance in research they support.

Table 2.3 Examples of Agency Enforcement Authorities

AGENCY	LEAD ENFORCEMENT AUTHORITY
Health and Human Services	Office for Human Research Protections
Central Intelligence Agency	Human Subject Research Panel
Department of Defense	Director, Defense Research and Engineering
Department of Veterans Affairs	Research Compliance Officers; Office of Research Oversight
National Aeronautics and Space Administration	Chief Health and Medical Officer

Sources: Department of Health and Human Services, Office of the Assistant Secretary for Health, Office for Human Research Protections. Program Description; Letter from V. Sue Bromley, Associate Deputy Director, CIA, to Amy Gutmann, Chair, Presidential Commission for the Study of Bioethical Issues. (May 16, 2011); DOD Directive 3216.02, Protection of Human Subjects and Adherence to Ethical Standards in DOD-Supported Research, sec. 4.10, March 25, 2002; VHA Handbook 1058.01: Research Compliance Reporting Requirements, para. 6; VHA Directive 1058, The Office of Research Oversight. www1.va.gov/vhapublications/ViewPublication.asp?pub_ID=1825. NASA. (2007).NASA Policy Directive: Protection of Human Research Subjects. June 14. Retrieved from http://nodis3.gsfc.nasa.gov/npg_img/N_PD_7100_008E_/N_PD_7100_008E__main.pdf.

Recent reviews of OHRP's enforcement actions show that noncompliance often involves violations related to the IRB initial review process and IRB approval of informed consent documents.[63] The identified violations were overwhelmingly procedural, with a much smaller percentage of the violations reflecting substantive concerns such as a failure to obtain informed consent (8 percent of violations).[64] These data offer a snapshot of the ways that institutions have failed to comply with regulations. Given the volume of research HHS supports, these results suggest that there is room for improvement, but that the incidence of serious harm or unethical treatment is fairly low.

Once a database of federally sponsored human subjects research is readily available, there are various methods that could be useful to assess investigators' specific understanding of ethical requirements and practices on the ground. For example, a project-by-project assessment could be undertaken using a sample of recent federally supported projects in order to determine how individuals and organizations conducting human subjects research currently apply federal regulations and international standards. Such an assessment could identify the practical ethical challenges facing researchers and the organizations that oversee them. Document review of agency IRB records, structured interviews with research team members and other stakeholders, as well as site visits, could bring far more specific information than is now readily available. Given limited information and time, the Commission did not undertake these actions, but it encourages future reviewers and policymakers to consider them.

Conclusion

Existing evidence suggests both that the rules governing federal research today adequately guard against abuses analogous to those perpetrated in Guatemala in the 1940s and that current regulations generally appear to protect people from avoidable harm or unethical treatment, insofar as is feasible given limited resources, no matter where U.S.-supported research occurs. This conclusion, set forth more fully below, is both consistent with and also qualified by the large, yet incomplete, set of information made available to the Commission in the time available to carry out its charge.

The Commission concludes, however, that improvement of the current system is both possible and desirable. In Chapter 3, the Commission describes those areas where improvements are warranted at this time. With these improvements, or response from policymakers to explain to the public as to why they are not justified, the abuses of the past can be left firmly in the past. There is no way to eradicate all risk of harm, particularly in some types of medical and translational research, but the current system, with the improvements suggested here, can become one in which we can all be more confident.

In sum, the Commission finds:

The current U.S. system provides substantial protections for the health, rights, and welfare of research subjects and, in general, serves to "protect people from harm or unethical treatment" when they volunteer to participate as subjects in scientific studies supported by the federal government. However, because of the currently limited ability of some governmental agencies to identify basic information about all of their human subjects research, the Commission cannot say that all federally funded research provides optimal protections against avoidable harms and unethical treatment. The Commission finds significant room for improvement in several areas where, for example, immediate changes can be made to increase accountability and thereby reduce the likelihood of harm or unethical treatment.

CHAPTER 3
Further Analysis and Recommendations

In several important ways, the current regulatory and professional system for the protection of human research subjects can and should be strengthened. Ethical standards, at least as reflected in current regulations, are sometimes seen as a barrier or burden rather than as an integral part of the web of respectful human relationships.[65] Society as a whole often lacks adequate information with which to hold researchers and research funders accountable for the work that they do. And, despite the fact that science itself is founded on the expansion and exchange of human knowledge, there are insufficient data available to assess the soundness of current human subject protections or their accessibility and responsiveness to stakeholders and the general public.

The Commission identified several areas for improvement, including:

1) increasing accountability;

2) helping those who are harmed as a result of research participation;

3) respecting equivalent protections of international partners;

4) promoting a culture of responsibility;

5) evaluating site selection and the justification for chosen study designs; and

6) engaging communities at all levels of research.

Some of these recommendations include specific suggestions for government action and others are directed more broadly to the community of research investigators, scientists, and others involved in protecting human subjects from harm and unethical treatment. The Commission also offers endorsement and comments on several of the proposed reforms published by HHS and OSTP in the ANPRM to reform the Common Rule issued in July 2011.[66]

The Commission's attempts to understand the full range of human subjects research conducted or supported by the federal government highlighted the difficulty of performing such an inquiry. Thus, recommendations are offered on increasing accountability of research programs. Review of existing regulations, guidelines, and professional standards; combined with testimony received about the challenges of linking procedural requirements to ethical

FURTHER ANALYSIS AND RECOMMENDATIONS   III

principles; underscored the need to focus on a variety of means to further cultivate and reinforce a culture of responsibility among researchers.

In examining the current system to ensure that subjects are protected from harm or unethical treatment, the Commission also identified a set of important ethical questions about study design and site selection that remain unsettled, yet possible to address in the context of particular research endeavors. Ethical considerations for the choice of study design include a range of issues, such as ascertaining to whom the benefits of the trial accrue, and determining whether the interests of the subjects and their communities are properly weighed and balanced in relation to the distribution of risks and benefits. Study design and methods cannot be justified by the importance of research alone. Certain study designs—such as comparing placebos to experimental interventions through a controlled trial—raise ethical concerns when existing alternative treatments can be effective in the communities where the research is intended to occur, unless a strictly defined set of conditions can be met under which comparison is justified. Engaging with relevant communities prior to beginning research may be one important step for researchers to show respect and sensitivity to cultural norms relevant to protecting human subjects.

The globalization of biomedical and behavioral research has at times outpaced the thoughtfulness with which decisions can or should be made about the selection of research sites, a decision that should never be simply a matter of efficiency or expediency. Ethical concerns inherent in moving research across international borders need further careful specification. Many issues are unlikely to be unique to international research; they also arise domestically when, for example, research is undertaken in resource-poor or culturally distinct settings. It is recognized that taking research to specific sites that are less expensive because of lower costs associated with care or recruitment, or because of the relative ease of identifying populations with the condition under study who might be treatment naïve does not necessarily mean the research is ethically unacceptable. But when the circumstances of selecting research sites suggest the possibility or appearance of exploitation and failure to respect individual human dignity or appropriate community interests, site selection—both domestic and international—needs to be examined more carefully than it has been at times in the past.[67]

Finally, it has long been agreed, as a matter of principle, that respectful treatment of those who volunteer as research subjects should include medical treatment for injury, however rare, without charge to subjects. While people may disagree on the extent to which medical expenses should be paid and by whom, the notion that research subjects should not be made worse off as a result of volunteering to advance science and society is generally uncontroversial. The just distribution of burdens suggests that individuals should not be forced to bear the costs of injuries they suffer to advance collective well-being. But here, as in much of life, the devil lies in the details. The Commission recognizes that there are variable policies and practices in place now to insulate subjects from bearing medical treatment costs and believes that the federal government should undertake further study to determine whether, and if so how, a government-wide policy could be developed.

The recommendations in this report are intended to enhance and strengthen the current system. They include some overarching issues applicable to research in any sector and several other issues that, while focused largely on details of health research, have important implications for how human subjects are protected. The Commission's recommendations, if implemented, will help to ensure that the well-being and rights of research subjects are protected, promote professionalism, better integrate ethical responsibilities among researchers, and serve to increase society's confidence that research volunteers are not subjected to avoidable harm or unethical treatment. In short, these recommendations offer tools to aid the practice of ethically sound research. There of course can be no guarantor of perfection or final word in this complex realm. Just as the President asked the Commission to examine our country's understanding and application of fundamental ethical principles for the protection of human subjects in research today, future generations will need to re-evaluate their contemporary application of these principles in the context of their own time.

Commission Recommendations

1. Improving Accountability

The need for greater accountability among all federal supporters of research emerged as a crucial, and imminently remediable, issue among the Commission's overarching concerns. Contemporary tools for public access to

FURTHER ANALYSIS AND RECOMMENDATIONS III

information offer mechanisms to enhance accountability. Expanded support for research into the methods and processes of human subjects protection offers the opportunity to further assure that human subjects are protected from harm or unethical treatment.

Science requires substantial societal investment, putting it in competition with other important activities that may contribute to the public good. Therefore, and especially for publicly funded research, the public has the right to accountability in the use and management of the resources allocated to the pursuit of scientific knowledge for the common good. Moreover, public scrutiny can offer the opportunity for responsible officials and scientists to gain a better understanding of the values and preferences of the public whose good they ultimately serve. As subject matter experts may sometimes fail to appreciate all implications of their work, substantive contributions by others, not directly engaged, may provide unexpected and positive contributions.

However, the rationale for accountability is not merely one rooted in the contingencies of public funding. Modern science is grounded in the enlightenment values of democratic participation and public demonstration, so that access to insights about the nature of the universe is not limited to only a small fraction of society. Over the last 200 years—and especially in recent decades—science has developed into an increasingly collaborative activity among teams that are often geographically distant, each taking responsibility for an aspect of a complex body of work. Peer review, which is an integral and longstanding practice in decisions about both funding and publication of research, is another element that crucially contributes to accountability and exemplifies the fact that science is a communal activity. Making research more transparent can help to improve its quality, reduce the chances for unnecessary duplication, and inform determinations of scientific merit and reasonable risk in human subjects research. Systems for making research information publicly available should take into account privacy and intellectual property concerns, and conform to applicable statutory and regulatory constraints.

Scientific and lay communities have increasingly advocated for public access to information about federally sponsored biomedical research activities. Insufficient access to research information allows studies and results to be hidden and can result in injuries to human subjects, wasted resources, and unethical exposure to unnecessary risk.[68] Federal law since 1997 has required public and online access to limited, basic information about clinical trials involving drugs for serious or life-threatening conditions, whether publicly or privately funded.[69] Congress expanded these requirements in 2007 to cover most clinical trials of drugs and devices and public access to basic results of these trials through the government website, ClinicalTrials.gov.[70] The International Federation of Pharmaceutical Manufacturers Associations and the State of Maine promulgated policies mandating public access to certain clinical trial information, and the World Health Organization developed a voluntary international clinical trial registry platform in which many countries participate.[71] Several countries maintain their own national registries a well.[72]

> **EXAMPLES OF CLINICAL TRIAL REGISTRIES AROUND THE WORLD**
>
> **Australia and New Zealand** – http://www.anzctr.org.au/
>
> **Brazil** – http://www.ensaiosclinicos.gov.br/
>
> **China** – http://www.chictr.org/en/
>
> **Cuba** – http://registroclinico.sld.cu/
>
> **European Union** – https://www.clinicaltrialsregister.eu/
>
> **Germany** – https://drks-neu.uniklinik-freiburg.de/drks_web/
>
> **India** – http://www.ctri.in/
>
> **Iran** – http://www.irct.ir/
>
> **International Standard Randomised Controlled Trial Number Register** – http://www.isrctn.org/
>
> **Japan** – http://rctportal.niph.go.jp/en/index
>
> **Netherlands** – http://www.trialregister.nl/trialreg/index.asp
>
> **Pan Africa** – http://www.pactr.org/
>
> **South Korea** – http://ncrc.cdc.go.kr/cris/index.jsp
>
> **Sri Lanka** – http://www.slctr.lk/
>
> Source: World Health Organization. (n.d.). Members of the WHO Registry Network. Retrieved from http://www.who.int/ictrp/network/primary/en/index.html.

The *Declaration of Helsinki* now states that "[e]very clinical trial must be registered in a publicly accessible database before recruitment of the first subject."[73] Federal law has also mandated online public access to basic information about federally funded grant and contract research. For example, USASpending.gov was implemented by the Federal Funding Accountability and Transparency Act to create a publicly accessible, online database of all federal contracts and grants over $25,000; cataloguing details such as the

FURTHER ANALYSIS AND RECOMMENDATIONS

funding agency, the recipient, project site, and a description of the project's subject matter.[74] Recovery.gov is a similar site that allows the public to track use of funding made available specifically under the American Recovery and Reinvestment Act.[75]

Despite these several and promising recent developments in making information about research publicly available, more can be done. Only "applicable" clinical trials must be registered in ClinicalTrials.gov, yet human subjects research encompasses a much broader range of study types. Biomedical research alone comprises first-in-human and other early phase trials, natural history studies, and physiological studies that are not among those required to be registered. Research with human subjects also includes, for example, social, behavioral, and economic research conducted by federal departments and agencies such as the Department of Education, the National Science Foundation, and the Department of Justice. Indeed, the National Science Foundation and Department of Education were among the top five federal supporters of human subjects research (by number of projects) in Fiscal Year 2010. (See Table I.1.) And, databases such as USASpending.gov and Recovery.gov track federal grants and contracts generally, not those related to research—much less human subjects research—specifically.

Other notable efforts to enhance accountability and access to information about human subjects research are also underway. For example, "Health Research Web," initiated by the Council on Health Research for Development, an international non-governmental organization whose primary objective is to strengthen research for health and innovation, compiles information on studies taking place around the world, as well as information about IRBs and research ethics committees, countries'

> **CLINICALTRIALS.GOV**
>
> ClinicalTrials.gov contains information about nearly all federally and privately funded clinical trials investigating new drugs or devices, except phase I drug and device trials and certain types of post-market surveillance. Registration of covered studies in ClinicalTrials.gov is mandated by statute.
>
> Publicly available data at ClinicalTrials.gov includes: the trial's sponsor; the title, description, and design of the study; locations where the study is conducted; and contact information for the trial facility. Clinicaltrials.gov does not collect funding information.
>
> Source: Food and Drug Administration Amendments Act § 801, 42 U.S.C. § 282 (j) (2007).

governance policies, and other useful data.[76] In 2008, Congress required that NIH-funded investigators submit their publications to PubMed Central, an electronic publicly accessible database.[77] And, in accordance with the America Competes Act, OSTP, in late 2011, requested public input on recommendations on approaches for "encouraging broad public access to unclassified digital data that result from federally funded scientific research" and "broad public access to the peer-reviewed scholarly publications that result from federally funded scientific research."[78]

Department- or agency-specific systems to track human subjects research are also available, though they vary widely. Some are centralized, publicly available databases. For example, the Department of Energy's Human Subjects Research Database provides a wide range of protocol-level information about Department of Energy-supported human subjects research, including funding for and number of subjects involved in specific studies.[79] NIH maintains RePORTER, a database listing information about all NIH grants and contracts such as funding amount, funding organization and location, and project abstract.[80] Other departments and agencies lack such systems. For example, as part of its review, the Commission learned that some agencies are unable to link information about human subjects research projects to funding information in any systematic way. Other agencies maintain numerous internal databases, but cannot relate or de-duplicate among them.

Comparability of existing agency-specific databases is also limited. The data elements contained in each database reflect individual agency priorities and resources. For example, NIH's RePORTER database, provides online access to award-level rather than project-level data, and includes information on all research—not just human subjects research. Therefore it can be difficult to readily distinguish information on specific human subjects studies.

In an effort to collect information on the scope, volume, and type of research funded by the federal government, the Commission requested from departments and agencies limited, project-specific information; such as study title, performance country (or countries), number of subjects (if available), and funding information; that it deemed necessary to assess the adequacy of protections in scientific studies supported by the federal government. For its Human Subjects Research Landscape Project, the Commission asked each

FURTHER ANALYSIS AND RECOMMENDATIONS III

department and agency covered by the Common Rule to provide data that they maintain and have readily available. Some elements were particularly difficult for many agencies to report, such as a study's number of subjects and whether the study was exempt from IRB review.[81] Each department and agency asked for data provided some information about its human subjects research portfolio to the Commission. Some were able to respond fully (or almost fully) with each of these variables, while others provided more limited information. (See Appendices I and II.)

As technical computing and data sharing capabilities have evolved, and the public is increasingly able to access information about the research that it supports, accountability in research can be increased. While access to information alone does not ensure that human subjects will be protected from harm or unethical treatment, making information related to human subjects research readily available across department and agency supporters will contribute to the protection of human subjects by enabling further scrutiny and accountability. Knowing where research occurs will enable interested members of the public to raise appropriate questions on, for example, site selection or research design. It will enable accrediting organizations and oversight officials to readily identify basic details about research within their jurisdiction.[82] And it will enable funders to more easily monitor their own portfolio to identify what work they do with human subjects and where more resources or attention may be needed.

The Panel also recommended that information on all research that is more than minimal risk be made publicly available through an online tool.[83] Given the existing requirements already in place for online access to information about many clinical trials posing more than minimal risk, the Commission does not single out this kind of research for special treatment. Rather, it recommends that a core, minimal data set on all federally funded research with human subjects should be made publicly available.

Existing systems—both agency-specific and government-wide—provide a foundation from which agencies can develop and improve their own systems to be utilized as publicly available databases. It may be particularly important that departments and agencies collect sufficiently detailed data for individual projects to be identified and categorized. Aggregate data do not provide information that would enable members of the public, including fellow researchers, to identify individual projects, understand where research may be duplicative (therefore imposing unnecessary risks on human subjects), or assess if research subjects are adequately protected. A core set of data elements about each project is necessary to provide a baseline description of federally supported human subjects research and changes over time, and to allow for compilation across departments and agencies. This core set of data elements should include at least: 1) title, 2) principal investigator and institutional affiliation, 3) location of research activities, and 4) funding source and funding amount.[84] Where possible, the nature of the research study or abstract of the study and number of participants should also be included. For clinical research, an "NCT" number, assigned to projects registered in ClinicalTrials.gov, if available, should be included to facilitate accountability by linking to other readily available information already maintained by the government.

The Commission considered whether a new, central, government-wide database of this information should be developed. Because of cost and capital investments required, the Commission recommends instead that each department and agency should make core data publicly available either by developing or improving upon its own systems to make information publicly available or release information through a trans-agency system. For agencies without existing systems today, data could be submitted to ClinicalTrials.gov or another existing online registry, provided the core data elements are available.[85] The Commission endorses registration and reporting results of all human clinical research including early phase studies and all privately funded research. Certainly it will be easier if the public can access as much information as possible through an existing and widely recognized mechanism such as ClinicalTrials.gov. Moreover, information submitters may find it easier to rely on this existing system rather than having to prepare information for multiple databases.

Creation of a central government web-based portal rather than a single government-wide database would enable department- or agency-level systems to be centrally accessible, leverage department- or agency-specific systems already in place, permit variability among agencies (e.g., differences in mission, research types, and recordkeeping practices), and provide opportunities for innovation and flexibility. A central web portal linking to each

FURTHER ANALYSIS AND RECOMMENDATIONS III

department or agency's system should be created and administered by OHRP or another designated agency. OHRP "provides leadership in the protection of the rights, welfare, and wellbeing of subjects involved in research conducted or supported by the U.S. Department of Health and Human Services,"[86] and also occupies a prominent role in human subjects protection government-wide.[87] Thus, maintenance of a central web portal for research information access is consonant with OHRP's mission. As above, however, recommendation of a portal mechanism is not intended to preclude the prospective development of a unified federal database if it is ultimately more cost-effective and efficient to do so and it is not intended to discourage use of available mechanisms that may make it easier for the public to obtain information efficiently.

Accordingly, the Commission recommends:

Recommendation 1: Improve Accountability through Public Access

To enhance public access to basic information about federal government-funded human subjects research, each department or agency that supports human subjects research should make publicly available a core set of data elements for their research programs—title, investigator, location, and funding—through their own systems or a trans-agency system. The Office for Human Research Protections or another designated central organizing agency should support and administer a central web-based portal linking to each departmental or agency system. This should not preclude the prospective development of a unified federal database that may ultimately be more cost-effective and efficient.

In addition to making core information available and more transparent for public scrutiny, the Commission finds that data are essential to determining whether current policies are meeting their intended goals. Contemporary federal policies regarding human subjects protection are generally based on sound ethical principles and many include procedural mechanisms that are intended to satisfy them. Determining the extent to which these procedures actually serve to protect human subjects depends, at least in part, upon empirical evidence. Today, there are limited data that demonstrate the effectiveness of particular procedures and no agreed upon metrics by which to measure them.[88]

Empirical and conceptual bioethics research synergistically play important roles in protecting human subjects. The need for empirical research in bioethics is well established. Descriptive data can inform and test normative work in bioethics.[89] It amplifies the force of conceptual bioethics activity, such as the development and delineation of ethical principles, by enabling such work to more closely map onto real-world situations and provide concrete solutions and recommendations. "[E]mpiric bioethics research serves [at least] three essential functions . . . : 1) debunking widely held but erroneous views; 2) assessing the importance of ethical concerns; and 3) facilitating the realization of certain ethical values."[90] The latter functions can help determine the extent to which contemporary research policies manifest abstract values and ethical principles as well as tangible human subject potections.

Recognizing the value of empirical research to advancing human subjects protection, Dr. Francis Collins, NIH Director, detailed for the Commission his agency's expanding research agenda in this area. Remarking that NIH Institutes and Centers have devoted roughly $50 million per year to research in bioethics, according to the most recent survey, Dr. Collins noted that the largest program, garnering $18 million per year, resides in the National Human Genome Research Institute, as its Ethical, Legal, and Social Implications Program. This program has been investigating the implications of genetic and genomic research for the last 17 years. During the two-year period following passage of the American Recovery and Reinvestment Act, NIH funded 21 bioethics Challenge Grants, which are noteworthy for the short time period and the especially competitive funding and review process. Set aside in the budget of the Office of the Director is $5 million for institutes that propose projects that fit the bioethics research and training agenda. A bioethics taskforce, comprised of representatives from 25 Institutes and Centers, has reviewed the

> "The near-term priorities have certainly emphasized the importance for mission-related bioethics initiatives and training initiatives. The longer-term goal is to try to do more to integrate bioethics into the full spectrum of biomedical research. Not having this as a separate discipline but one that is fully connected with what is going on in the laboratories and the clinics."
>
> Dr. Francis Collins, Director, NIH, (2011). Bioethics Research at the NIH. Presentation to PCSBI, February 28, 2011. Available at: http://bioethics.gov/cms/node/187.

results of NIH investment in bioethics research and is now entering a second phase and developing a strategic plan for the future of bioethics research at NIH. This strategic plan is certain to include initiatives to "integrate bioethics into the full spectrum of biomedical research," making it part and parcel of the scientific process instead of a separate discipline.[91]

Still, there remains a dearth of knowledge about the actual efficacy of human subjects protections. Given this, the Commission recommends that the federal government support an expanded operational research agenda to study the effectiveness of human subjects protections. Research and evaluation designed to identify and measure protections afforded by Common Rule procedures could also inform understanding and implementation of the Common Rule's equivalent protections provision, evaluate the effectiveness of implementing guidelines for community engagement, and assist in clarifying the justifications underlying research-site selection, all of which are areas the Commission believes can be improved (see *Respecting Equivalent Protections, Promoting Community Engagement,* and *Justifying Site Selection,* below). Accordingly, the Commission recommends that the federal government continue to support and consider expanding its support for systematic assessment of current human subjects protection standards.

Recommendation 2: Improve Accountability through Expanded Research

To evaluate the effectiveness of procedural standards embedded in current human subjects protection regulations, the federal government should support the development of systematic approaches to assess the effectiveness of human subjects protections and should expand support for research related to ethical and social consideration of human subjects protection.

Taken together, these recommendations aim to make more data publicly available in a coordinated way, enable periodic independent evaluation of the scope and volume of federal human subjects research, and support ongoing empirical study of human subjects protection standards. With them, the public will be better informed about federal research with human subjects, domestically as well as internationally, and will have greater opportunity to hold researchers and other officials engaged in the research process accountable for the work that they do.

2. Treating and Compensating for Research-Related Injury

Ethical Justification for Compensation

Those who sponsor or engage in human subjects research have an ethical obligation to protect people who volunteer as research subjects. Just as health services researchers speak of both primary and secondary forms of prevention, the Commission understands the duty to provide "protection" to encompass two important kinds of duties—primary and secondary protection.[92]

Lifeguards, for instance, have a duty to protect swimmers in both a primary and a secondary sense. They are obliged to protect swimmers from drowning risks, *and* to rescue those who begin to drown. Similarly, researchers should protect subjects from exposure to undue risk *and* limit or reverse the harm subjects may experience as a result of their participation in the research by assuring the provision of appropriate medical care.

The duty to protect only concerns risks that can be avoided or remedied. One can only protect against risks that can be reasonably foreseen and avoided or risks due to states of affairs that can be identified and modified. Some risks cannot be avoided, and there can be no duty to protect against such risks in a primary sense. Some harms cannot be remedied, and there can be no duty to protect against such harms in a secondary sense.

In thinking about whether there is an ethical justification for providing treatment or some form of compensation for the medical costs of research-related injuries, two important facts about the risks of research must be considered:

- First, for the most part, the benefits of such research redound to the common good, while the risks are almost entirely borne by the subjects.[93]

- Second, some risks of human subjects research are unavoidable and often not readily foreseeable. This unforeseeable nature of some of the risk is part of the ethical justification for undertaking human subjects research in the first place.

Many have argued that there is a duty to provide medical care for subjects harmed by their participation in research.[94] This Commission is not the first to study the question of compensation for injuries incurred in the course of research participation. The Presidential Commission for the Study of Ethical Problems in Medicine and Biomedical and Behavioral Research in 1982 and NBAC in 2001 explored this issue and called on the government to study the feasibility and

FURTHER ANALYSIS AND RECOMMENDATIONS III

need for requiring treatment or compensation for medical costs in the United States.[95] The Institute of Medicine in 2002, also recommended that "organizations conducting research should compensate any research participant who is injured as a direct result of participating in research, without regard to fault."

Over the last several decades, almost all other developed nations, and many transnational standard-setting bodies, have instituted policies to require researchers or sponsors to provide treatment or compensation for treatment for research subjects' injuries.[96] (See Table 3.1.) The Commission's Panel recommended to the Commission that the United States should establish a system to assure compensation for the medical care of subjects harmed in the course of biomedical research.[97]

Table 3.1 International Requirements for Treatment of Research-Related Injury

COUNTRY	POLICY
Belgium	Requires subjects receive treatment for injury and compensation for death.
Denmark	Establishes fund ensuring treatment for injury and additional compensation for pain and suffering, spouse or partner, loss of earnings, loss of earning capacity, loss of dependency for children, and injury to a person's feelings or reputation.
Finland	Establishes fund ensuring subjects receive treatment for injury and additional compensation for pain and suffering, disability, and economic loss.
Uganda	Requires that subjects receive treatment for injury and compensation for any resultant impairment, disability, or handicap.
Brazil	Requires that subjects receive comprehensive medical care for injury to the physical, psychic, moral, intellectual, social, cultural, or spiritual dimensions of the human subject.
GUIDANCE DOCUMENT	**RECOMMENDATION**
CIOMS, International Ethical Guidelines for Biomedical Research Involving Human Subjects	Recommends that subjects receive both treatment for injury, recommends compensation for resulting disabilities.
International Conference on Harmonisation, Guideline for Good Clinical Practice	Recommends that subjects receive treatment for injury, and recommends compensation for resulting injuries in accordance with applicable regulatory requirements.
Institute of Medicine, Responsible Research: A Systems Approach to Protecting Research Participants	Recommends that subjects receive compensation that includes at least the cost of medical care and rehabilitation.

Sources: Belgium: Annex No. 1 Law of May 7, 2004 Concerning Experiments on the Human Person. Denmark: The Danish Liability for Damages Act (2005, amended 2006 and 2007). Finland: Pharmaceutical Injuries Insurance: General Terms and Conditions (2007). Uganda: Ugandan National Council for Science and Technology. (2007). *National Guidelines for Research Involving Humans as Research Participant*. Kampala: UNCST. Retrieved from http://www.uncst.go.ug/dmdocuments/Guideline,%20Human%20Subjects%20Guidelines%20Marc.pdf; Hungary: Act XCV of 2005 on Medicinal Products for Human Use and on the Amendment of Other Regulations Related to Medicinal Products. (2005). Brazil: Rules on Research Involving Human Subjects. (2003). Council for International Organizations and Medical Sciences and World Health Organization. (2002). *International Ethical Guidelines for Biomedical Research involving Human Subjects*. Geneva: World Health Organization. ICH/GCP: International Conference on Harmonisation. (1996). Guideline for good clinical practice E6(R1). June 10. Retrieved from http://www.ich.org/fileadmin/Public_Web_Site/ICH_Products/Guidelines/Efficacy/E6_R1/Step4/E6_R1__Guideline.pdf; Institute of Medicine. (2002). *Responsible Research: A Systems Approach to Protecting Research Participants*. Washington, D.C.: National Academies Press.

The primary argument in favor of providing care or costs for care of research-related injuries involves the principles of justice and fairness. Those who agree to take part in human subjects research accept the risks of this research, and place their bodies and sometimes even their lives on the line. For taking on the unavoidable risks of bodily injury that are inherent in such research, subjects often receive little or no monetary benefit and often no promise of any direct therapeutic benefit, particularly in early phases of clinical research. Society benefits from research subjects' acceptance of these risks, so it seems fair that they are protected from some of the ameliorable harms that they may sustain as a result of their participation. This is sometimes described as an argument against "free-riders."[98] Those who benefit have an obligation in fairness toward those who experience harms.

Unintended harm is inevitable in the course of human subjects research. Human subjects who are harmed as a consequence of participation in research should not individually bear the costs of medical care for such harms. This is true whether the harm results from a foreseen risk about which they are informed as part of the informed consent process or from a wholly unforeseen risk. Treatment or compensation for the cost of medical care is appropriate as an act of benevolent regard for individuals' willingness to participate in an enterprise of important benefit to the public.

> "I think compensation is the great leveler, and … I think if the victim is harmed through no fault of his or her and through no fault of the researcher, I think there has got to be [compensation]."
>
> Mr. Feinberg drew a contrast with the current tort system, noting that a compensation program, if implemented, should be an "efficient streamlined system that can reach resolution in months, not years," noting that the current system is "often not very efficient."
>
> Kenneth R. Feinberg, J.D., Administrator of the Gulf Coast Claims Facility and Special Master of the September 11 Victim Compensation Fund. (2011). Presentation to the PCSBI, November 17. Available at: http://bioethics.gov/cms/meeting-seven.

The primary argument against this view has been the assertion that there is a freely undertaken assumption of risk on the part of subjects volunteering for human research. If they understand the risks and benefits and still freely consent to participate, so the argument goes, then they have no claim, in justice, for compensation for any harms that befall them as a result of their participation in the research. This argument is steeped in legal history and expressed in the doctrine of *volenti non fit injuria* (those

FURTHER ANALYSIS AND RECOMMENDATIONS

> "We need human subjects for the simple reason you can't try new therapies on lots of people that you haven't tested. So we have to have human subjects. And there's going to be an unavoidable, ineliminable burden that this places on some human beings. . . . [W]e have to be very careful that we spread this risk fairly[.]" But, "doing it without a safety net is not required. You can do the science and provide a safety net. . . . [T]here's no excuse for imposing that on subjects whereas there is a perfectly good excuse for imposing the burden of being a research subject, namely you can't do research without it."
>
> Daniel Wikler, Ph.D., Mary B. Saltonstall Professor of Population Ethics, Professor of Ethics and Population Health, Department of Global Health and Population, Harvard University. (2011). Compensation for Research-Related Injury. Presentation to the PCSBI, November 17, 2011. Available at: http://bioethics.gov/cms/node/391.

who consent cannot claim injury).[99] Because society permits research subjects to consent freely and accept the risks of research, then it is claimed that society should permit subjects to consent freely and accept the risk of doing so without expectation of free medical care or compensation for any harm caused by their participation.[100]

There are several arguments against the view that a human subject's free and informed consent to participate in research qualifies as assumption of all risks related to research participation. This conclusion rests on a fallacious conflation of avoidable and unavoidable risks. Society does not permit subjects to consent to any and all research risks, but requires instead a prior determination by an IRB that foreseeable risks are minimized or eliminated.[101] IRBs carry out this duty of primary protection by prohibiting research that involves undue risk, regardless of whether an individual may freely consent to accept those risks. IRBs permit subjects to consent and accept unavoidable risks when the benefits of the research outweigh the perceived risks of harm. But many research risks are not reasonably foreseeable or cannot be avoided; this is one of the very reasons why research is conducted. While the nature of research at times requires the acceptance of unavoidable physical and psychological risks, to which society permits subjects to freely expose themselves, this fact does not mandate that society must permit subjects also to accept avoidable risks before, during, or after research. Among the risks that can be avoided is the cost of medical treatment for unavoidable injuries. Protecting human subjects from bearing the costs of medical treatment for illness or injury from participation, that is identifiable and ameliorable through medical care, comes under the category of secondary protection and the duty to assist such subjects arises again.[102]

A just society can protect persons from undue risks. A society that values individual freedom and autonomy can consistently determine that some arrangements are inherently unfair and unjust. It can also determine that some parties are particularly vulnerable and not in the best position to make individual determinations that would protect them from undue risk. Society makes such a determination in preventing subjects from enrolling in clinical trials that are too risky, and charges IRBs with making that determination on a routine basis. If research subjects are entitled to such primary protection, they should also be extended secondary protection if they are harmed as a result of their participation.

The paradigmatic example of this sort of research circumstance is the typical Phase I trial of a new but traditional cytotoxic oncologic agent that holds promise in animal studies but has not yet been tried in human beings. The purpose of such a study is to determine toxicity and tolerability in human beings. The subjects enrolling in such trials are typically patients with advanced malignancies for which all standard therapeutic options have failed. Such studies pose risk and hold very little promise of therapeutic benefit for those who volunteer as subjects. The risks are largely unforeseeable and unavoidable. These subjects are vulnerable, often desperate, holding out hope for individual benefit even though informed that the probability of individual benefit is small. Even if an individual's own motives are not altogether altruistic, the results of such trials nonetheless redound to the common good. Even negative trials contribute to our common scientific understanding. To say to such persons that they are volunteering for the common good but that they must bear full financial responsibility for medical care needed to treat any harms that ensue as a direct result of their participation seems grossly unfair.

Nor does the argument for compensation depend on the untenable assumption that research is conducted always and only to benefit the general public. The argument instead is that voluntary acceptance of risk by human subjects, which advances the interest of the biomedical research enterprise, warrants benevolent and just responses. Even when the motives of the investigators and sponsors are not solely to advance the common good, the fact remains that the research enterprise does redound to the common good regardless of the motives of the investigators. Moreover, human subjects are explicitly recruited to volunteer for research, even in the for-profit sector, under the assumption that their efforts are at least partly intended to advance the common good.

FURTHER ANALYSIS AND RECOMMENDATIONS

Still another objection to compensating research subjects might be that the presumption that subjects are volunteers is naïve: some research today, it may be claimed, comes closer to the market conditions of laborers and employers freely contracting, with both primarily motivated by personal gain. This is the model of the research subject as "wage-earner,"[103] which some commentators have explicitly endorsed.[104] Adopting the wage-earner model full-throttle, however, actually enhances the case for compensation on the basis of justice, rather than undermining it. If research subjects are employees, and employees in a dangerous job, then how can they be justly excluded from a form of worker's compensation that is available to other employees in other industries? Clearly any harm caused to their health by virtue of their participation in the research would be "work-related" injuries that ought to entitle them to compensation.

Whichever model one chooses for understanding the relationship between human research subjects, investigators, and sponsors, justice argues for a system that assures that research subjects who suffer medical, dental, or psychiatric harms directly caused by their participation in research ought not to bear the costs of treatment for these harms all by themselves.

Moral considerations other than justice also argue for treatment or compensation for treatment of research-related injuries. To the extent that human subjects research is biomedical and conducted by physicians, nurses, dentists, clinical psychologists, and other health care professionals, then professional ethical commitments to the principles of beneficence and non-maleficence also support a system of compensation for those harmed by research. One might argue that clinician-investigators who know that at least some individuals will inevitably be harmed by their human subjects research ought not to engage in such research unless they can be assured that there is a system in place to care for those harmed by research so that their duties of beneficence and non-maleficence can be fulfilled.

Finally, general utility also argues for such a system. Potential human research subjects may be more likely to agree to serve if they know they will be taken care of in the event that they are harmed as a direct result of their participation. In an era in which the recruitment of adequate numbers of research subjects continues to be a major challenge, this could be significant.

Further, if there is a growing commitment toward harmonization of policies and recognition of equivalent protections, the fact that federally sponsored human subjects research differs substantially in its policy towards compensation for human subjects who are harmed in the course of research could prove a significant barrier. Domestically, private sponsors of human subjects research have voluntarily committed to assuring such protections to research subjects, and internationally, many other countries require this. Harmonization and equivalent protections could be well-served were the federal government either to prove that such a system already exists in a rigorous (even if inchoate and patchwork) fashion, or to adopt a more formal and transparently comprehensive system of compensation.

The Commission concludes that ethics requires that subjects harmed in the course of human subjects research ought not individually bear the costs of care required to treat qualified harms resulting directly from that research.[105] Such a conclusion does not, however, specify what an optimal system to carry out this ethical mandate would look like. Some serious considerations include the scope of any possible coverage, the delineation of qualified harms, mechanisms for determination of causation and qualification, relation to the tort system, the need for any special public or private insurance, and how the current nonsystematic approach to this issue functions in practice. A distinction also needs to be made between compensation for the costs of needed care resulting from harms due to participation in morally justifiable research and reparations for unethical research.

Reparations for Unethical Research

The foregoing discussion concerns treatment or compensation to treat harms sustained as a result of participation in trials that are ethically justified. The Commission finds that it is important to distinguish this from possible reparations for unethical and exploitative research.

The above discussion focused on compensation (or restitution) through providing medical care or the funds to remunerate participants in human subjects research for the direct costs of medical care for injuries that have proximately resulted from their participation in research. The Commission argues that these costs should not be borne by the subjects themselves. Under the presumption that these injuries were the result of ethically acceptable research and could not reasonably have been prevented but instead resulted directly from the uncertainties associated with the experimental intervention (and which justified undertaking the research project in the first place), the investigators cannot be held ethically blameworthy for such injuries.

Thus, treatment or compensation/restitution for the costs of treatment, is justified by distributive or corrective justice and by duties of beneficence. It is also justified by the primary duty of professional caregivers to do no harm. For example, someone who enrolls in an ethically approved and conducted clinical trial, who suffers from pneumonia caused by the trial drug but who lacks health insurance to pay for treatment, therefore could merit treatment or compensation for treatment costs. Likewise, someone currently enrolled in an ethically designed and ethically conducted U.S.-sponsored trial carried out overseas who develops a severe side effect and needs health care but has no independent means to obtain treatment, such as insurance for health care, could likewise merit treatment or compensation for the costs to treat the research-related harm.[106]

Reparation, by contrast, calls for acknowledgment of wrongdoing and contrition, along with actual or symbolic repayments for wrongdoing. The Commission uses the term reparation in this context to describe the expression of regret for wrongs done to victims of unethical human subjects research. Such individuals need not have been harmed physically or psychologically, and may not even be alive. Likewise, the individual perpetrators of the unethical research may no longer be alive to be held directly accountable.

However, once unethical research has come to light, the institutional sponsors of such unethical research bear some responsibility to make amends for past institutional wrongdoing or that of their former agents. As such, a formal apology, compensation for any identifiable living individuals harmed by such research, or symbolic gestures of contrition such as the establishment of charitable foundations or institutions related to the future prevention of such harms may also be appropriate in some cases.[107]

Compensation might be part of reparation. For example, an individual who was enrolled in the infamous Public Health Service's "Tuskegee Study of Untreated Syphilis in the Negro Male" (Tuskegee Syphilis Study) from 1932-1972, and now, still alive, needs treatment for tertiary syphilis, might deserve compensation as part of an overall program of reparation.[108] Reparation, however, need not include compensation. For example, in 1997,

President William Jefferson Clinton apologized to the families of those who were enrolled in the Tuskegee Syphilis Study, and the United States provided funds for the establishment of the Tuskegee University National Center for Bioethics as an act of reparation.[109] President Obama's swift apology to the people of Guatemala for the U.S. Government's sponsorship of the 1946-1948 STD studies was similarly an important act of reparation, or moral repair.

Designing a System of Compensation

The Commission recommends that the federal government undertake a careful assessment to address how best to satisfy the ethical obligation to compensate individuals who suffer research-related injuries as a result of volunteering in a federally funded study. The nature and extent of injury, the type of research in which the injury is occurring, and the costs of injury to subjects, investigators, and society have not been systematically studied. When the scope and nature of compensation for research-related injuries is determined, it will be possible to address adequately the practical questions associated with treatment and compensation.

The questions to be addressed in such an assessment include: To what extent do established (and emerging) public and private health insurance programs contribute to compensating individuals for research-related injuries? What types of injuries are compensated? How is it best to establish causal links between research protocols and medical problems? How should research subjects in foreign countries be compensated? Should there be limits placed on the time, amounts, and categories of compensation?

The Commission was informed and advised about the advantages and disadvantages of various means by which the current system offers treatment or compensation for treatment for some research subjects. These means include: civil tort liability, institutional self-insurance and commercial insurance, individual health insurance, government benefit programs like Medicare, and direct treatment paid or provided by agencies directly (and as a term and condition of a research award). As a general matter, the tort system works well for many types of intentional and accidentally caused injuries. Some alternatives that selectively pre-empt the tort system, such as the National Vaccine Injury Compensation Program (which the Panel proposed as a model for consideration), have been established to take into account the particular

FURTHER ANALYSIS AND RECOMMENDATIONS

circumstances under which compensation is due.[110] The Commission, accordingly, advises of the importance of carefully considering the variation among federally sponsored human subjects research in order to determine the optimal system for compensation.[111]

Today, several federal departments and agencies provide treatment or compensation for treatment when research subjects are injured. The Departments of Defense and Veterans Affairs both provide care for research-related injuries. The Department of Veterans Affairs' regulations require that it provides care for all research-related injuries, even in those studies considered minimal risk.[112] The Department of Defense provides health care services from military treatment facilities for subjects injured in the course of research, but no compensation (i.e., payment) for injuries.[113] The NIH Clinical Center provides short-term care but no long-term care or financial compensation for injury.[114] These programs essentially "self-insure" for treatment or costs of treatment.[115] Personal insurance or government programs such as Medicare also pay the costs for treatment for some injuries arising from research.[116] (See Table 3.2.)

Table 3.2 U.S. Treatment/Compensation for Treatment Methods

INSTITUTION	TYPE OF POLICY
NIH Clinical Center	Provides short-term care during the trial, but no long term care or financial compensation.
NASA	Provides compensation for injuries arising from intramural research through its worker's compensation system; directs principal investigators for extramural research to provide compensation through "insurance, worker's compensation, or the like."
Medicare	Medicare covers "reasonable and necessary items and services used to diagnose and treat complications arising from participation in *all* clinical trials." Medicare serves as a secondary payer for these costs.
University of Washington	Self-insurance: no-fault program that provides up to $10,000 for out-of-pocket costs and write-off of care provided at University of Washington Medicine.
Private clinical trial insurance providers (e.g.: RJ Ahmann Company)	Covers a number of different types of liability, including: general liability, latent injury liability, and incidental medical malpractice liability.

Sources: National Institutes of Health. (2006). *Sheet 6—Guidelines for Writing Informed Consent Documents*, retrieved from http://ohsr.od.nih.gov/info/sheet6.html (accessed November 23, 2011); NASA. (2004). NASA Procedural Requirement 7100.1 – Protection of Human Research Subjects, secs. 9.1.4, 11.6, and Appendix B, Revalidated July 7, 2008; Centers for Medicare and Medicaid Services. Medicare Coverage—Clinical Trials: Final National Coverage Decision. Retrieved from: https://www.cms.gov/clinicalTrialPolicies/Downloads/finalnationalcoverage.pdf; Centers for Medicare and Medicaid Services. (2010, May 26). Clinical trials and liability insurance (including self-insurance), no-fault insurance, and workers' compensation. Retrieved from: http://www.cms.gov/MandatoryInsRep/Downloads/AlertClinicalTrailsNGHP.pdf; Moe, K.E., Director and Assistant Vice Provost For Research, University of Washington. (2011). University of Washington Human Subjects Assistance Program. Presentation to the PCSBI, November 17. Retrieved from http://bioethics.gov/cms/meeting-seven; RJ Ahmann Company. (n.d.). What You Need to Know About Insuring Your Clinical Trials. Retrieved from http://www.rja.com/clinical-trial-insurance-liability-minneapolis-mn/.

Some academic centers that receive federal funds for research with human subjects also provide treatment and cover medical costs for research-related injuries. These are established through institutional self-insurance and commercial insurance schemes. For example, since 1972 the University of Washington health care system has administered a university-wide system of treatment and compensation for treatment of research-related injuries. This system, self-funded through the institution's operating budget, provides up to $10,000 for out-of-pocket expenses and for treatment at University of Washington facilities.[117]

Sometimes component programs of federal departments and agencies require awardees to provide for treatment of research-related injuries or carry insurance to cover treatment costs. The National Aeronautics and Space Administration directs principal investigators seeking funding for research conducted outside of its facilities to provide compensation for injury "by means of insurance, worker's compensation, or the like" and may fail to approve research because such a provision is not included in a protocol.[118] The National Human Environmental Effects Research Laboratory (funded by the Environmental Protection Agency) promises to provide up to $5000 to cover costs for treatment of research-related injuries.[119] Costs for insurance are paid by the government as part of research awards.[120]

FDA guidance directs research investigators as part of their duties to protect the rights, safety, and welfare of subjects and to provide both "reasonable medical care…for medical problems arising during participation in [a] trial that are, or could be, related to the study intervention," and "reasonable access to needed medical care, either by the investigator or by another identified, qualified individual (e.g., when the investigator is unavailable, or when specialized care is needed)."[121]

Most industry-based clinical research sponsors carry insurance to compensate individuals injured in research trials. In discussion and written comments to the Commission, representatives from PhRMA and the Biotechnology Industry Organization, trade groups representing pharmaceutical and biotechnology companies respectively, explained that good business practice, as well as legal obligations, require their member companies to carry insurance to cover the costs of research-related injuries.[122]

FURTHER ANALYSIS AND RECOMMENDATIONS   III

No authoritative sources exist by which to determine whether all individuals for whom compensation for research-related injuries is warranted are both adequately and quickly compensated for their medical care. Consequently, in order to fulfill the President's aim of ensuring that all human subjects are adequately protected in federally supported research, the Commission finds that further study into any shortfalls of the existing system and the utility of alternative systems of compensation is necessary. The Commission recognizes that it would be unwise to alter or supplant the country's current approach to compensation for research-related injuries without first gauging the nature and scope of harms that remain unaddressed. A careful and timely study, however, will enable the government to address the practical questions associated with whether or how to develop a better overall system or a supplementary set of approaches to the current system that optimally assures compensation.

For example, the government may decide that a study of alternative approaches to compensation should take into account considerations such as deterrence, loss spreading, and internalization of risk.[123] These considerations raise questions about where the burden of compensation is best placed. For example, how much (if any) of the burden of compensation should be placed on research sponsors to create and sustain incentives for ensuring the highest possible level of protection of human research subjects consistently with preserving the incentives and capacity for research? Are researchers themselves in the best position to internalize some or all of those costs, that is, to price some or all of the costs of compensation—including insurance and administrative costs—into their research programs? Because researchers play an essential role in advancing biomedical research and discovery, an effective compensation system should not *unnecessarily* burden them or impede their ability to undertake novel research programs that advance scientific progress and discovery. The Commission is recommending a study to help the government determine what constitutes a manageable cost of compensation and who should bear these costs.

Alternative, complementary, or supplementary models to the status quo that merit consideration range from case-by-case compensation by insured or self-insured research sponsors and institutions, to a more centralized governmental system of compensation for research subjects (analogous to the current National Vaccine Injury Compensation program), to the creation of wholly new institutions or system-wide requirements (that specify what constitutes adequate standards of compensation). Some compensation programs preempt state tort remedies, while others do not. Statutes such as the National Childhood Vaccine Injury Act and the legislation that established the Workers Compensation fund preempt state tort remedies.[124] The study that the Commission recommends should explicitly consider whether federal preemption of state tort remedies is necessary or desirable for any new (or revised) compensation model, whose goal is to ensure coverage of all qualified research-related injuries. The study should also consider whether a single and predictable model of compensation is more or less burdensome to researchers and secure for subjects than a piecemeal approach, the overarching ethical goal being to ensure that research participants who are injured by research—whether by foreseeable or wholly unexpected injuries—receive adequate medical care for their injuries.

The Commission's deliberations found that National Childhood Vaccine Injury Act model or other strict liability models may not be appropriate for human subjects research, and could be difficult to apply in a wide range of research settings. Many retrospective federal systems of compensation—such as the September 11th Victim Compensation Fund—are not appropriate models: they have been responses to disastrous events that called for a coordinated *unique* national response.[125] The National Childhood Vaccine Injury Act, which is a prospective model of compensation, was also a response to concerns of vaccine manufacturers about the risk-benefit ratio of producing vaccines in light of the substantial civil liability they were then facing and could continue to face.[126]

The Commission thinks it important to recognize the limits of these models in the course of considering the optimal means (which are likely to be plural, rather than singular) of ensuring appropriate compensation for research-related harms. A study should take into account the differences between the conditions of being recruited and volunteering for the kind of experimental research that may later result in harm and being a subject who uses an approved, manufactured product that later results in harm. Similarly, designing a system of retrospective compensation for harms sustained by

FURTHER ANALYSIS AND RECOMMENDATIONS

many people during a single involuntary event, for example the September 11th Victim Compensation Fund, differs from designing a prospective program for compensating people harmed by having volunteered to assist the common good.

Comparing possible clinical compensation schemes to vaccine compensation schemes may provide useful insight. For example, clinical research often occurs in environments where research subjects are asked to take medicines or other interventions outside of the clinical setting. Vaccines are administered in a clinical setting where one can be certain that the vaccine was administered, when it was given, and to what individual. A relatively small number of signature adverse events arise from vaccines, and when one of those adverse events occurs, the vaccine more likely than not caused the injury. The side effects that arise from clinical research trials may not be as limited or as predictable. There are a large number of research protocols and a wide range of medical interventions, each of which may produce a variety of adverse events.

Any system of compensation needs to define a standard or standards for when an injury would become a treatable or compensable event.[127] Such standards could turn on the nature or the severity of the injury, and whether side effects, including those that follow an effective therapeutic intervention—such as anticipated or unanticipated side effects—should constitute a compensable event. Moreover, a standard or standards should articulate what is needed to show that the research caused the injury.

Recommendation 3: Treating and Compensating for Research-Related Injury

Because subjects harmed in the course of human research should not individually bear the costs of care required to treat harms resulting directly from that research, the federal government, through the Office of Science and Technology Policy or the Department of Health and Human Services, should move expeditiously to study the issue of research-related injuries to determine if there is a need for a national system of compensation or treatment for research-related injuries. If so, the Department of Health and Human Services, as the primary funder of biomedical research, should conduct a pilot study to evaluate possible program mechanisms.

Recommendation 4: Treating and Compensating for Research-Related Injury Follow Up

The Commission recognizes that previous presidentially appointed bioethics commissions and other duly appointed advisory bodies have made similar recommendations regarding compensation or treatment for research-related injuries; yet no clear response by the federal government has been issued. Therefore, the federal government, through the Office of Science and Technology Policy or the Department of Health and Human Services, should publicly release reasons for changing or maintaining the status quo.

3. Creating a Culture of Responsibility: Human Research Protections as Professional Standards

The most fundamental obligation of research involving human subjects is to protect the rights and welfare of individuals who offer themselves for the good of both science and society and, in some cases, for the hope of personal benefit. As persons, research subjects possess an inviolability that rules out treating them as mere means to the ends of others, including others who may be suffering from a disease or in need of medical care. But, in helping to advance research, human subjects sometimes place themselves in a position of informational asymmetry, where they must rely on the expertise and wisdom of researchers, reviewers, funding institutions, and, at times, their own physicians to ensure that a research study in which they enroll is designed and deployed with their rights and welfare in mind.

Respecting the interests of subjects is often more complicated than it might appear. Despite comprehensive (and what some describe as overly burdensome) regulations, a number of competing interests may make it difficult for researchers to recognize or exercise objective ethical judgment in practice.[128] For example, in clinical trial design researchers may struggle with trade-offs between high quality science—which offers the possibility of developing future means to prevent, treat, or reverse disease—and the immediate best interests of subjects.[129] So too, financial and nonfinancial conflicts of interest, or the appearance of these conflicts, have received increasing attention in the last decade. When manifest, they can lead to harm and unethical treatment and jeopardize participant and public trust in the research system.[130]

FURTHER ANALYSIS AND RECOMMENDATIONS III

As early as 1966, physician Henry Beecher introduced the idea that the best protection for subjects is the enlightened and ethically sensitive conscience of the investigator. In addition to the requirement of informed consent, Beecher described "the presence of an intelligent, informed, conscientious, compassionate, responsible investigator" as offering the best protection for research subjects.[131] In his view, a system of external regulation was unnecessary because properly educated and virtuous clinician-researchers would know to do the right thing based on their keenly honed internal compasses. In spite of Beecher's preference for an enlightened conscience over the sometimes-heavy hand of regulation, a number of highly publicized research scandals, some identified by Beecher himself, in the last century demonstrated the need for external regulation.[132]

While a separate source of oversight in the form of regulations and enforcement is an indispensable complement to a culture of virtuous investigators, the optimal system for human subjects protection is a dual system of external regulatory checks and internal embodiment of appropriate professional norms, such as respect for persons. In developing such a system, care must be taken to ensure that external safeguards are not so onerous as to dampen the desire and commitment to develop an internal ethical identity. Ideally, researchers should view the norms of research ethics as a legitimate and necessary part of their mission as scientists. In this regard, the principle of regulatory parsimony, which the Commission developed more fully its report, *New Directions: The Ethics of Synthetic Biology and Emerging Technologies*, must be acknowledged in concert with regulatory necessity.[133] Only by balancing compulsory and voluntary checks can a lively personal sense of professional ethics develop.

Ethics in a professional practice should be seen as strict standards that are simultaneously self-imposed by the relevant profession and expected by the society it serves, similar to any other practice standards or standards of care. Rather than a matter of only regulatory compliance, professional ethics are an integral part of what it means to be a profession in the public service. Professional standards correlate with ethical duties toward subjects and with the privileges of office and cannot be waived or ignored for expediency, convenience, perceived interests of the many at the expense of the few, or the allegedly superior demands of science. Researchers must view compliance as a shared duty stemming not only, or even primarily, from regulation, but instead from their collective professional responsibility to the subjects of their research and the public that supports their research mission.

Today, many sources exist to describe useful professional norms. For example, PhRMA, of which many of the major pharmaceutical research and biotechnology companies are members, has released a code entitled PhRMA's *Principles on Conduct of Clinical Trials and Communication of Clinical Trial Results*.[134] The goal of this code is to uniformly assist member companies in the ethical conduct of clinical trials. Many member companies have also adopted their own code of conduct to promote professional standards.[135]

In order for researchers to understand, internalize, and embrace critical ethical requirements for research with human subjects, these obligations need to be seen as strict professional standards that accompany and justify the privileges of office. Throughout their professional training, researchers should learn and adopt these standards as a condition of acceptance into the professional discipline entrusted to undertake research with fellow human beings.[136] Funding agencies and research institutions should commit to creative, flexible, and innovative educational approaches, such as case studies combined with rigorous ethical analysis, undergraduate and graduate level mentoring, and even senior researcher "coaching," both in science and in ethical treatment of subjects by master researchers.[137] A shift towards teamwork and away from individual blame in research institutions may also facilitate the discussions that will promote a culture of responsibility.[138] The aim should be to both intellectually engage researchers and instill a sense of ethical responsibility as an integral part of professionalism. Success in achieving this aim could avert the resentment of learning "modules" that many well-motivated researchers currently view as an insulting waste of time.[139]

> "Education regarding research and research participation is critically important for all stakeholders including institutional leadership, IRBs, investigators and research staff, policymakers, sponsors, research subjects and the general public."
>
> Letter from Barbara E. Bierer, Chair, Secretary's Advisory Committee on Human Research Protections (SACHRP), to Kathleen Sebelius, Secretary of Health and Human Services. (August 5, 2011). Retrieved from http://www.hhs.gov/ohrp/sachrp/commsec/commentspcsbi.pdf.pdf.

As important as it is to teach rigorously the ethics of protecting human subjects to professionals, it is at least as important to teach the increasingly important and well-researched subject of bioethics—including research ethics—at all levels, including the

FURTHER ANALYSIS AND RECOMMENDATIONS III

undergraduate level. Bioethics is a universally important subject, fully consonant with a liberal arts and sciences education, and as such it should not be taught first, let alone only, at the professional-school level. Human subjects are drawn from the general citizenry, not only or mainly from the ranks of medical professionals and researchers. Citizens will benefit from knowing their rights as human subjects, and accountability will be better served if citizens are educated to hold accountable the researchers, IRBs, funders, and others in the research enterprise. The Commission recommends the development and expansion of rigorous courses in bioethics and human subjects research tailored to students at different levels of education. A proven effective method of teaching ethics is to combine scholarship on the subject with case studies—such as the STD experiments in Guatemala—to engage students in deliberations about what they would and should do in these and analogous circumstances.[140] Like the Panel, the Commission strongly recommends that ethics education play an increasingly central role in advancing research ethics.[141]

The Commission recommends that specific training in research ethics be focused not simply on the system of regulations but also on the ethical principles animating them. Professional societies, universities, and accrediting organizations need to promote standards not as legal burdens relegated to compliance departments but as expectations enforced by the community of scientists as well as oversight officials. Regulators should explicate the ethical rationale for requirements reflecting these standards in order to foster a strong sense of professional responsibility.

Accordingly, the Commission offers the following recommendations:

Recommendation 5: Make the Ethical Underpinnings of Regulations More Explicit

To promote a better understanding of the context and rationale for applicable regulatory requirements, the Department of Health and Human Services or the Office of Science and Technology Policy should ensure that the ethical underpinnings of regulations are made explicit. This goal is also instrumental to the current effort to enhance protections while reducing burden through reform of the Common Rule and related Food and Drug Administration regulations. (See *Promoting Current Federal Reform Efforts* below.) Following the principle of regulatory parsimony, regulatory provisions should be rationalized so that fundamental, core ethical standards are clearly articulated.

Recommendation 6: Amend the Common Rule to Address Investigator Responsibilities

The Common Rule should be revised to include a section directly addressing the responsibilities of investigators. Doing so would bring it into harmony with the Food and Drug Administration regulations for clinical research and international standards that make the obligations of individual researchers more explicit, and contribute to building a stronger culture of responsibility among investigators.

Recommendation 7: Expand Ethics Discourse and Education

To ensure the ethical design and conduct of human subjects research, universities, professional societies, licensing bodies, and journals should adopt more effective ways of integrating a lively understanding of personal responsibility into professional research practice. Rigorous courses in bioethics and human subjects research at the undergraduate as well as graduate and professional levels should be developed and expanded to include ongoing engagement and case reviews for investigators at all levels of experience.

4. Respecting Equivalent Protections

Clinical research has increasingly become a global enterprise. The number of privately sponsored clinical trials being conducted around the world has grown dramatically, as has the number of multinational and collaborative research projects sponsored or supported by the federal government.[142] Most, but not all, human subjects research supported by the federal government is subject to the same regulatory requirements—either the Common Rule and/ or FDA regulations—regardless of where it is conducted. Research collaborators and partners in other countries funded or supported by Common Rule agencies must file an assurance that they will comply with Common Rule requirements. At the same time, the regulations delineated in the Common Rule have long permitted U.S. departments and agencies supporting or conducting research to recognize and accept procedures from foreign countries that may differ from those delineated in U.S. regulations as long as they provide "protections that are at least equivalent" to those in the Common Rule. Yet U.S. departments and agencies have rarely, if ever, exercised the

FURTHER ANALYSIS AND RECOMMENDATIONS

authority to accept any foreign country's procedures as equivalent. In some cases U.S. procedural requirements even conflict with individual country requirements, and controversy remains regarding the processes or criteria to determine equivalence of protections.[143]

The Common Rule applies to "research conducted, supported, or otherwise subject to regulation by the Federal Government outside the United States";[144] and states that the policy "does not affect any foreign laws or regulations which may otherwise be applicable and which provide additional protections to human subjects of research."[145] The Common Rule also gives authority to a department or agency to approve the substitution of procedures utilized by foreign countries that differ from those delineated in the Common Rule if the department or agency determines that those procedures "afford protections that are at least equivalent" to those in the Common Rule.[146] Since the time when U.S. federal regulations and the Common Rule were first written, many countries have developed their own regulations and laws regarding the protection of human research subjects and some have even adopted laws or guidelines based on the International Conference on Harmonization Good Clinical Practice Guidelines or on the Common Rule itself. In some cases, other countries' laws and regulations are even more extensive than those found in the Common Rule.

One way of demonstrating respect for foreign collaborators and partners in research, a goal articulated by the Panel, is to develop and employ a process for determining when protections delineated in foreign laws and regulations are equivalent to U.S. regulations.[147] In addition, as noted by SACHRP, allowing recognition of equivalent protections could reduce the burden on U.S. IRBs without reducing, and perhaps even enhancing, human subject protections.[148] Furthermore, this process will facilitate ongoing international dialogue between U.S. and international bodies, the importance of which the Panel also highlighted in its recommendations to the Commission.[149]

> "Recognizing equivalent protections would minimize the problem of U.S. insistence on procedural standards that may not offer more effective ethical safeguards for human subjects, or that may preclude research in countries where it could improve public health."
>
> The International Research Panel of the PCSBI (2011, September). Research Across Borders: Proceedings of the International Research Panel of the PCSBI. Washington, D.C.: PCSBI, p. 9.

The issue of whether and how to recognize "equivalent protections" offered by non-U.S. collaborators and researchers is not new. NBAC recommended in 2001 that the U.S. government "identify a set of procedural criteria and a process for determining whether the human participants protection system of a host country or a particular host country institution has achieved all the substantive ethical protections."[150] Also in 2001, the U.S. Office of the Inspector General, citing the increasing volume of international collaborative research, recommended that OHRP address how to "better assess whether other nations' laws and practices afford equivalent protections to those that apply to human subjects participating in clinical trials in the U.S."[151]

Two years later, in 2003, an internal HHS working group made several useful recommendations and proposed a framework for developing criteria to determine equivalent protections. The first step in its framework is to articulate the specific protections embodied in the Common Rule. The working group noted that the Common Rule is primarily procedural and does not articulate either the ethical basis for its required procedures or the actual protections that its procedures provide. The working group identified seven protections it thought the Common Rule does afford, including that it: 1) establishes expectations of ethical conduct and due diligence in review and performance of research within the institution; 2) ensures adequate independence and authority of the IRB/Research Ethics Committee; 3) protects from biased and arbitrary decisions in research ethics review; 4) ensures sufficient quality and comprehensiveness of research ethics review; 5) ensures review and oversight are commensurate with risk and the vulnerability of the study population; 6) protects from unnecessary or unjustified risk throughout the study; and 7) ensures voluntary participation after adequate disclosure of study information. Importantly, the working group observed that "the protections embodied in the [Common Rule]…generally represent broad and complex aims, each of which might be satisfied in a variety of different ways through a variety of procedures."[152]

Despite this 2003 HHS report and a 2005 *Federal Register* notice calling for public comment on the proposals,[153] OHRP has not finalized or employed criteria or process for determining equivalent protections. In fact, a 2006 *Federal Register* Notice on Interpretation of Assurance Requirements reiterated

FURTHER ANALYSIS AND RECOMMENDATIONS

the requirement that all who receive U.S. federal research funds, including those outside the United States, follow the exact procedures delineated in the Common Rule.[154] United Kingdom government officials in 2007 formally requested a determination from HHS regarding the equivalence of protections afforded by United Kingdom procedures for human subjects protection, and other countries are also interested in gaining recognition of "equivalent protections" for their system.[155] Yet to date OHRP has not formally recognized any country's protections as equivalent.

The federal government does recognize the adequacy of some foreign human subjects protections standards as part of its program to license new drugs and devices for use in the United States. FDA regulations for clinical investigations supporting new drug or device marketing applications specify requirements, similar to those found in the Common Rule, for informed consent, IRB review, and other standards to protect human subjects. FDA accepts data from foreign studies that comply with certain international standards for human subjects protection (such as those that abide by "good clinical practice," the *Declaration of Helsinki*, or certain host country regulations), rather than with FDA's exact regulatory procedures.[156] Thus, the development of a system for recognizing equivalent protections under the Common Rule can draw from the experiences of FDA.[157]

Given this history, and the prominent way that recognizing equivalent protections demonstrates respect for communities (a safeguard discussed further below), the Commission recommends:

Recommendation 8: Respect Equivalent Protections

The federal government, through the Office for Human Research Protections, should adopt or revise the 2003 Health and Human Services Equivalent Protections Working Group's articulation of the protections afforded by the specific procedural requirements of the Common Rule. It should use these requirements to develop a process for evaluating requests from foreign governments and other non-U.S. institutions to determine if their laws, regulations, and procedures can be recognized as providing equivalent protections to research subjects.

5. Promoting Community Engagement

The Panel directed the Commission's attention to the value of community engagement and the demonstration of respect for cultural differences that are compatible with the ethical conduct of human subjects research. The term "community engagement" means many things. It includes but is not limited to: community based participatory research, which is a term of art for studies designed to engage communities and researchers as equal partners, studying health issues in community settings, using communities or community entities as a unit of study or unit of randomization, engaging communities in defining research priorities, engaging communities in shaping a particular study to their needs, as well as community "consent," community advisory boards, community outreach for recruitment, and community monitoring of study progress.[158]

In the United States, community advisory boards include, for example, the Framingham Heart Study Ethics Advisory Board, which enables participants and community members from the town of Framingham, Massachusetts, to advise researchers conducting the decades old NIH-funded epidemiological study on proposed research design and methods.[159] Community advisory boards are required for each research site in the HIV Vaccine Trials Network, a group of NIH-funded researchers working across the globe to find a vaccine for HIV led by researchers at the University of Washington.[160] And recent guidance from the Centers for Disease Control and Prevention, the Health Resources and Services Administration, and NIH contains comprehensive discussion of how public health researchers and others can engage community partners in their work.[161]

The values underlying calls for community engagement are applicable to research conducted both domestically and abroad. Effective community engagement provides an additional layer of safeguard by providing the community with opportunities to more thoroughly weigh and accept or reject the risks and benefits of research activities, discover possible implications of research that might have unintended consequences to host communities, and independently evaluate the effectiveness of research protections.[162] Interactive and ongoing dialogue between the communities engaged in research and the research team allows for the integration of community norms, beliefs, customs, and cultural sensitivities with the research activities.

FURTHER ANALYSIS AND RECOMMENDATIONS   III

Community engagement should provide ongoing two-way communication between community partners with research funders, teams, and national regulatory authorities. As discussed by the Panel, such communication serves as a local mechanism of accountability for the researchers and can also create a mutual sense of partnership that facilitates realizing the goals of the research project. In the event that communication reveals conflicting ethical values among partners regarding how to conduct the research, the partners can engage in a mutually acceptable process of conflict resolution.

Over the past few decades, there has been a growing international consensus on the principles underlying the ethical conduct of biomedical research. Although there remains some variation in their content, internationally recognized documents converge on several themes that are generally compatible with the idea of accommodating community norms in human subjects research, but only to the extent that those norms do not conflict with the principles that are essential to protecting individual subjects and otherwise making human subjects research ethical.[163] To this point, the Panel recognized, however, that "researchers cannot—and should not—accept uncritically everything that a community recommends or requests."[164] For example, several international documents admit the possibility that it may be both necessary and desirable for researchers to use a mechanism other than a written document to obtain individual informed consent in some communities, but they rule out substitution of the collective consent of a community, community leader, or a spouse for the consent of any adult who is capable of giving his or her individual consent.[165]

Valuing community engagement during human subjects research around the globe is a relatively recent phenomenon. This is seen especially in international guidance documents. Although the topic is tangentially discussed in the Council for International Organizations of Medical Sciences (CIOMS) 2002 *International Ethical Guidelines for Biomedical Research Involving Human Subjects*, CIOMS focuses its guidelines on responsiveness to the health needs and priorities of host countries, and the ethical acceptability of the study for the host country requiring "a thorough understanding of a community's customs and traditions."[166] The *Declaration of Helsinki* also includes provisions regarding the health needs and priorities of host countries, but no mention is explicitly made of community engagement standards.[167]

> "Why does this matter? Well, it matters not just for the conduct of a single trial. It matters for the ways trials can take place in a long-term process with communities over many different types of research endeavors. It really comes down to 'how do we create the trust and respect for the research process that researchers and clinicians have, but communities often don't for lack of their input and engagement throughout the process.'"
>
> Mitchell Warren, Executive Director, AVAC: Global Advocacy for HIV Prevention. (2011). Presentation to PCSBI, August 30. Retrieved from http://bioethics.gov/cms/node/319.

Still, these principles are beginning to emerge with more force in transnational documents. The 2007 Joint United Nations Programme on HIV/AIDS UNAIDS and World Health Organization (WHO) guidance document, *Ethical Considerations in Biomedical HIV Preventative Trials*, and the companion UNAIDS/AVAC *Good Participatory Practice Guidelines*[168] explicitly include community engagement principles. Nor has the lack of international guidance stopped the incorporation of community engagement principles in a piecemeal fashion by individual entities. For example, in addition to U.S. examples cited above, research ethics bodies in Australia and Canada have created documents outlining the importance of community engagement to ethical human subjects research, specifically focusing on aboriginal and other minority communities.[169]

Despite a dearth of explicit formulations for community engagement by international organizations, proposals for a common framework for community engagement in biomedical research have appeared in the literature over the past decade.[170] Still, there remains a lack of sufficiently detailed and extensive empirical information on the successful implementation of community engagement strategies for biomedical research and even less information for research in other spheres.[171]

In 2011, UNAIDS and AVAC published a new edition of their guidelines to provide trial funders, sponsors, and researchers with systematic guidance on how to effectively engage with stakeholders on the design and conduct of biomedical HIV prevention trials.[172] This publication provides a roadmap to implement community engagement based on principles of respect, mutual understanding, integrity, transparency, accountability, and community stakeholder autonomy.

FURTHER ANALYSIS AND RECOMMENDATIONS III

The UNAIDS/AVAC *Good Participatory Practice Guidelines* call for execution of community engagement through a set of activities, including use of community advisory boards, the development of community communication processes, discussion of post-trial access to beneficial research products or outcomes, and access to care for discovered medical illnesses. (See Figure 3.1.) These guidelines, however, were developed specifically for HIV prevention research rather than for human subjects research more generally. They should be prospectively evaluated for applicability to human subjects research beyond HIV prevention.

Layers of Biomedical HIV Prevention Trial Stakeholders

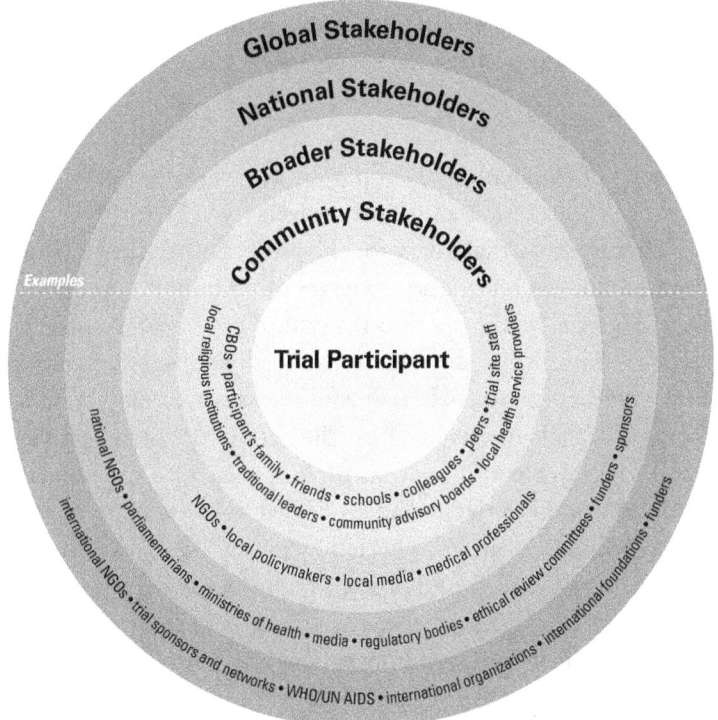

Source: UNAIDS/AVAC. (2011). *Good Participatory Practice Guidelines for Biomedical HIV Prevention Trials.* UNAIDS: Geneva. Retrieved from http://www.avac.org/gpp (accessed November 21, 2011).

Figure 3.1

To further develop operational guidelines for the protection and ethical treatment of human subjects through the means of community engagement, the Commission recommends that:

Recommendation 9: Promote Community Engagement

The federal government, through the Office for Human Research Protections and authorized research funders, should support further evaluation and specification of the Joint United Nations Programme on HIV/AIDS and the AVAC *Good Participatory Practice Guidelines* **with the aim of providing a standardized framework for those community engagement practices that would further the protection and ethical treatment of human subjects in all areas of research. Research should be conducted to prospectively evaluate the effectiveness of this framework and strengthen it after it is developed.**

6. Justifying Site Selection

Careful selection of research sites is important for two sets of ethical considerations. First, the ethical criteria for how subjects must be treated narrows the selection of sites to those that allow for the protection and ethical treatment of subjects. Second, site selection implicates other important considerations of fairness and justice. With regard to the ethical treatment of subjects, as the *Belmont Report* recognizes, "selection of research subjects needs to be scrutinized in order to determine whether some classes are being systematically selected because of their easy availability, their compromised position, or their manipulability, rather than for reasons directly related to the problem being studied."[173] With regard to the second set of ethical considerations, the report continued, "whenever research supported by public funds leads to the development of therapeutic devices and procedures, justice demands that these not provide advantages only to those who can afford them and that such research should not unduly involve persons from groups unlikely to be the beneficiaries of subsequent applications of research."[174]

Certain domestic and international settings present particular concern about possible exploitation of human subjects. Alan Wertheimer has observed that explicit concern for "exploitation" in research is pronounced in cases involving vulnerable populations and studies in low- and middle-income countries.[175] U.S. history contains examples of the failure to treat human subjects ethically in no small part because subjects were chosen for their vulnerability

FURTHER ANALYSIS AND RECOMMENDATIONS III

rather than for the sake of potentially benefiting them or members of their community. One prominent example where the site and the subjects selected for research signaled a willingness to exploit vulnerable and poor members of a minority population for the purported benefit of others was the Tuskegee Syphilis Study. In retrospect, some defenders of this research argued that subjects received some benefits through annual medical check-ups and minor health treatments otherwise unavailable to them, but indubitably these "benefits," standing alone, were far outweighed by the total neglect to treat the syphilis[176] and by the pervasive deceptions that defined this study.[177]

The STD experiments funded by the Public Health Service in Guatemala during the 1940s are another prominent example of how sites may be chosen for the vulnerability of available subjects, in this case, prisoners, psychiatric patients, sex workers, and conscripted soldiers. The U.S. researchers exploited the easy availability, compromised position, and manipulability of these Guatemalan subjects, and they benefited from the willingness of some Guatemalan officials and researchers to cooperate with and support the experiments.

> "The Guatemala research targeted some of the most vulnerable groups in any society (prisoners, conscripted soldiers, institutionalized psychiatric patients, and children), and also was conducted in an underdeveloped country with pervasive social inequalities that exacerbated their vulnerabilities. Such populations are given special protections in modern society because of their limited abilities to protect their own interests…In the Guatemala experiments the most vulnerable populations appear to have been targeted specifically because of their inability to protect themselves or to have others represent their interests."
>
> PCSBI. (2011, September). "Ethically Impossible" STD Research in Guatemala from 1946-1948. Washington, D.C.: PCSBI.

Decisions about research site selection involve consideration of many factors related to scientific opportunities and possibilities, as well as ethical considerations. When conducted according to the highest ethical and scientific standards, human subjects research across national borders can serve important scientific ends and yield life-saving and life-enhancing benefits for participating subjects and the communities in which they live.[178] An expanding number of publicly and privately funded research projects are multinational, involving numerous collaborating researchers and institutions, and are taking place in hundreds of sites in multiple countries; therefore, attention to the ethics of site selection is more important than ever. Some reasons for site selection are worrisome—for example, choosing a site because of the availability of vulnerable, treatment naïve subjects, lower regulatory burdens, and limited plans for making the experimental treatment available to its population. Similarly, choosing domestic or international sites where systems for protecting human subjects are not in place, are insufficient or inexperienced, or allow for the conduct of research that would not pass scrutiny elsewhere elevate concerns about exploitation and the possibility of unethical treatment.

The Commission strongly affirms that the same ethical principles that apply to domestic research should also be applicable on the international front. We should assess globally based research, including research in developing countries, by the same high ethical standards that we apply to domestic research and research in developed countries. Three longstanding and widely accepted principles relevant to the ethics of human subjects research are 1) do not treat people as mere means to the ends of others by doing research without their consent, 2) treat all individuals fairly and with respect, and 3) do not subject people to harm or the risk of harm, even with their consent, unless the risk is reasonable and there is a proportionate humanitarian benefit to be obtained.[179] These three fundamental principles do not change depending on the location of a study, the type of research, or the funding source. Nor are these principles of ethical research exclusively applicable to human subjects research in medicine or health. Because the most prominent examples of past abuses are found in biomedical research, and scholars' attention to site selection focuses on this area, the Commission directs this discussion of site selection to health-related research. However, the principled position staked out here can and should be extended beyond biomedical research.

The primary goal of biomedical research is the production of generalizable knowledge that will elucidate information about human health and illness and facilitate the design and application of preventive, diagnostic, or therapeutic interventions to meet the needs of future patients. Individual subjects of the research themselves may benefit by receiving medical attention in a study, and communities may benefit through the development of their health care infrastructures or if the fruits borne of a study are made available to them. Myriad risks comprise the burdens of research, including risks to physical health, mental health, privacy, confidentiality, social status, and economic productivity. Research that places a significant burden on subjects and/or their communities can constitute ethically impermissible exploitation unless the risks are offset by an adequate level of benefit, including type and quantity of benefit. Research in low-income countries raises particular concern that communities with limited access to needed health care will accept the risks of research but derive benefits that are disproportionately low in relation to the risks involved—or that they will be enrolled in research to answer questions that will only benefit those in richer countries.[180]

These concerns lead to the Commission's second set of reasons for calling attention to the issue of site selection: minimizing exploitation and promoting fairness and justice. One proposed strategy for minimizing the potential of exploitation when research is done in domestic or international low-income communities is to ensure that the proposed study is responsive to the health needs and priorities of the local community. Several international codes of ethics incorporate this broad criterion. For example, the CIOMS' 2002 *International Ethical Guidelines for Biomedical Research Involving Human Subjects* provide that, when conducting research in resource-limited areas, sponsors and investigators are responsible for responding to the "health needs and the priorities" of the subject population as well as ensuring that they can benefit from the research.[181] When research involves vulnerable or disadvantaged groups, the *Declaration of Helsinki* similarly holds that the justifiability of the research hinges on its responsiveness to local health needs and priorities.[182]

Generally speaking, responding to the health needs and offering benefits to a population may be an appropriate way to minimize the possibility of exploitation and promote fairness. The *Universal Declaration on Bioethics and Human Rights* (UDBHR), adopted by the General Conference of UNESCO in 2005, also emphasizes the importance of responsiveness to host communities, stating that "transnational health research should be responsive to the needs of host countries, and the importance of research contributing to the alleviation of urgent global health problems should be recognized" (Article 21). NBAC recommended that a developing country should only be selected as a research site when the proposed study responds to the host country's health needs (Recommendation 1.3).[183] Yet the concept of responsiveness is highly abstract and needs further specification before it can be operationalized; further, many more specific operational definitions that can be gleaned from the literature on responsiveness have been subject to strong counterarguments.[184] Prevalence of disease, for example, may be important in evaluating responsiveness, though it surely is insufficient as an indicator of burden of disease. Less common and even less burdensome diseases may be important health needs amenable to answers through good research. Understanding responsiveness as requiring research to respond to health priorities may provide a disincentive to research other important health problems. In this way, it could readily, albeit unintentionally, let the perfect—approving only ethically sound research that serves the highest health priorities of a poor community or country—become the enemy of the good—approving all ethically sound research that serves some health need.[185]

How to first effectively define and then implement the criterion of responsiveness remains an unsettled issue among scholars and practitioners.[186] Who has the capacity and legitimate authority to define what research is adequately responsive to a community, and how best to prioritize that community's health needs, are two critically important questions that remain in need of further careful examination. Precisely how any developed criteria of responsiveness to health needs apply to both publicly and privately funded research is yet a third question that cries out for further consideration. What is clear, however, with regard to the ethics of site selection is that all human subjects research should be performed in sites where the researchers are both willing and able to conduct their research in a way that protects human subjects from avoidable harm and unethical treatment. Enforcing this ethical mandate will also go a long way toward minimizing, even if not eliminating, the threat of unjustifiable exploitation.

Another important consideration in site selection, which enables ethical research to be conducted in under-served communities both domestically and internationally, embraces the values of well-functioning collaborations and well-established infrastructure for conducting ethical research and high-quality science, as well as assurance that the site has or can obtain the necessary infrastructure, resources, and oversight structure to ensure that research subjects will be treated ethically. With ample capacity to conduct research, robust oversight and enforcement mechanisms, and a strong ethical framework for human subjects protection comes the expectation in international research that the host country is well positioned to participate in scientifically valid research and adequately protect subjects in that research.

FURTHER ANALYSIS AND RECOMMENDATIONS   III

In considering the adequacy of infrastructure and ethical oversight across national lines, many questions arise, including: Are potential research subjects available and can they be fairly selected from the local population? Does the design of the trial match the capabilities of the possible site, in terms of facilities, staffing, and other assets to execute the study or can capacity be built to enable the conduct of the study without unreasonable diversion of limited resources? Is the site selection compatible with ensuring that protections for human subjects are equivalent or superior to those in the country originating the research, including mechanisms for evaluating and ensuring an acceptable balance of risks and benefits for the subjects, independent review of the trial design, informed consent, and fair compensation for research-related injury?

Careful examination of site selection is critical to ensuring that subjects in research are protected from avoidable harm or unethical treatment. The duty to ensure ethical site selection lies first with the researchers who conduct research. Generally, they are closest to the ground and able to judge conditions in the host community and country. Sponsors and funders too retain responsibility and should not initiate or approve research in locations where ethical site selection cannot be assured. Funders and sponsors may consider relying on determinations of duly constituted and independent IRBs and their equivalent bodies internationally. For example, a "just in time" certification, of the sort some agencies currently employ for institutional certification of IRB review could be developed to address the adequacy of site selection as well.[187] But funders and sponsors should not allow IRB or ethics committee representations to override or replace their own responsibility. In many cases, IRBs will have little influence over site selection, as study site decisions may be made far in advance of IRB review. Each actor in the funding, conduct, and oversight of human subjects research can, and should, exercise their discretion and authority to consider the ethics of site selection.

Recommendation 10: Ensure Capacity to Protect Human Subjects

Funders of research should determine that researchers and the sites that they propose to select for their research have the capacity—or can achieve the capacity contemporaneously with the conduct of the research—to support protection of all human subjects.

In addition to the absolute requirement to choose research sites where human subjects can and will be protected and treated ethically (and the implied prohibition on not choosing sites that cannot ensure this protection), other important considerations in site selection include how responsive to the health needs of a larger community a research project must be once human subjects are assured protection and ethical treatment. This issue raises complex questions of distributive justice in a non-ideal world where incentives to do publicly or privately funded human subjects research are often not well aligned with the locations that have the greatest needs for such research.[188] Recognizing that some of these questions remain unsettled and knowing that they will benefit from further research and analysis, the Commission concludes that the government should lead an effort to focus on them and develop more concrete guidance for researchers, funders, and IRBs.

Recommendation 11: Evaluate Responsiveness to Local Needs as a Condition for Ethical Site Selection

The federal government, through the Office for Human Research Protections and federal funding agencies, should develop and evaluate justifications and operational criteria for ethical site selection, taking into consideration the extent to which site selection can and should respond to the needs of a broader community or communities. The Office for Human Reseach Protections should produce, and other agencies should consider developing, guidance for investigators.

7. Ensuring Ethical Study Design

The scientific design of human subjects research, especially clinical research, becomes a subject of ethical concern when rigorous design elements (e.g., placebo controls, assignment to treatment arms, randomization, and blinding) raise the question of whether scientific advance is being placed ahead of human subjects protection.[189] For example, when subjects are assigned to a particular treatment or control group to receive an intervention known or strongly suspected to be inferior to other treatments available in other arms of the study or existing outside the study altogether, critics raise doubts about whether the rights or well-being of these subjects are compromised. Several controversial cases in the last 20 years have polarized the medical community into sharply opposed camps on this issue.[190]

FURTHER ANALYSIS AND RECOMMENDATIONS   III

> **DIFFERING VIEWS**
>
> *"Placebo Orthodoxy"*
>
> Comparing an experimental drug to a placebo ensures scientific validity and accurate measurements of drug efficacy. Thus, it is ethical to use placebos as controls, even when alternative, approved therapies exist.
>
> *"Active-Control Orthodoxy"*
>
> Placebos are never ethical if an alternative, approved therapy exists, and the best available alternative must be used as the control.
>
> *"Middle Ground"*
>
> A placebo-controlled trial can sometimes be considered ethical if certain methodological and ethical standards are met. If these standards cannot be met, then the use of placebos in a clinical trial is unethical.
>
> Emanuel, E.J., and F.G. Miller. (2001). The Ethics of Placebo-Controlled Trials—A Middle Ground. *New England Journal of Medicine* 345(12), 915-919.

Among the most discussed of these cases was a series of trials undertaken in the 1990s to gauge the safety and effectiveness of a short course of AZT (an antiretroviral drug) in preventing the perinatal infection of children with HIV. The active arm receiving AZT was compared to a control arm receiving a placebo (i.e., an inactive agent). The proponents of these trials contended that a placebo control was necessary in order to develop efficiently a scientifically validated, safe, and effective prophylaxis against HIV infection in children and noted that the use of the known effective method was not feasible in the test locations.[191] Critics argued strenuously that other designs (e.g., involving active controls and equivalency studies) were both scientifically reliable and ethically mandatory. They argued that those who designed, approved, and carried out such placebo-controlled trials were responsible for hundreds of preventable deaths.[192] Similar debates arise when human subjects are engaged in controversial procedures, for example "washout periods" that involve withdrawal of current medications for such serious conditions as depression or psychosis in order to avoid confusion with the effects of new drugs.[193] Also controversial, "challenge studies" involve intentional exposure to pharmacologic agents or circumstances to induce disease or symptoms for further study.[194]

Regarding the ethics of placebo-controlled clinical trials, differences remain and disagreements continue, but consensus is emerging.[195] In particular, Ezekiel Emanuel and Franklin Miller have proposed a "middle ground" for ethical clinical research,[196] largely supported by this report. Many ethicists and ethical codes have concluded that placebo-controlled trials can be ethical, provided certain criteria are met, including (but not limited to) these conditions: that withholding a proven treatment will cause only minimal harm to the subjects; that using an established intervention would not yield reliable results; and that the study is responsive to the needs of the host country. NBAC and CIOMS, for example, suggest giving the control arm of the trial the "worldwide best" treatment available as a default rule, with exceptions permitting a lower standard based on factors such as necessity, potential harm, affordability, and relevance of the research question to the host community.[197] The current *Declaration of Helsinki*, however, specifies that new interventions should be tested against the "best current proven" intervention, restricting the permissibility of placebo-controlled trials.[198]

In highlighting this emerging consensus, the Commission begins with the maxim that in research "good ethics begins with good science."[199] Every clinical trial starts either within a context of controversy, wherein investigators disagree as to the relative merits of, for example, new drug A and old drug B; or a context of uncertainty wherein investigators do not know which drug is better. Clinical trials are designed specifically to help resolve such controversy and uncertainty. If they are poorly designed and lack scientific rigor, studies cannot contribute to the resolution of such questions and the medical community will potentially ignore the results. Scientifically flawed studies have no social value, and if they have no social value there is no benefit to exposing subjects to any risk through participation. Ethical research must therefore be based upon a firm foundation of good (i.e., rigorous) science.

The unfettered demand for scientific and methodological rigor still can raise ethical concerns about whether subjects are being treated with sufficient deference to their status as moral agents worthy of equal concern and respect, persons should never be reduced to "mere means" for the benefit of others. For example, designing a placebo-controlled trial for a life-threatening illness such as HIV/AIDS might be thought to violate "clinical equipoise,"[200] which would ethically necessitate a genuine controversy or uncertainty among the expert medical community about the net benefit of each intervention being compared (including a placebo/no treatment at all). With convincing evidence and consensus within the medical community that, for example, drug A is better than drug B, then a design giving some human subjects drug A and others drug B, under the principle of clinical equipoise, would be unethical.[201]

FURTHER ANALYSIS AND RECOMMENDATIONS   III

Different commentators describe the nature of this perceived ethical flaw in different ways.[202] Some argue that researchers who knowingly withhold the best-proven treatment from subjects violate their "therapeutic obligation," which is grounded in the physician-patient relationship.[203] Others, owing at least in part to the fact that much human subjects research is not conducted by physicians (e.g., social scientists), locate the source of this duty elsewhere, such as in the subjects' status as equal and inviolable moral agents.[204] For example, critics of the perinatal AZT trials argued that a placebo-controlled trial exposed pregnant women and their children to an excessive risk of harm or death.[205] They charged too that when citizens of rich and prosperous countries receive the best-proven treatments, but people in lower and middle income nations receive "short course" treatments or placebos in clinical trials, this creates a morally pernicious "double standard."[206] This concern becomes even more problematic when the research is not "responsive" to the health needs of the host country.[207]

Towards a Middle Ground on Study Design

In the last section, the Commission discussed some ethical considerations that go into selecting research sites which must be considered in designing studies as well.[208] Furthermore, the Commission offers the following specific criteria for evaluating methods and risks of study design.

(1) Treatment Standards

As a matter of biological effectiveness and safety, treatments proven best for populations in the developed world are not always best for populations in the developing world. For example, genetic differences between population groups may make the best-proven treatment generally unsuitable for some, and variability in underlying health status may make some populations more vulnerable to side effects than others. Also, differences in the medical and logistical infrastructure of a developing host country may render the effective deployment of a best-proven intervention difficult or even impossible in practice. For example, the best possible comparator drug for the control arm might require refrigeration unavailable in a country lacking reliable electrical power, or the drug might need to be delivered intravenously in a country without the medical resources necessary for such delivery.[209]

The standard of care available to subjects in a control group need not be the best-proven, especially when there is uncertainty about whether such a standard would be best for the local population under study. Thus, in spite of the rhetorical appeal of the "no double standards" objection, it has become increasingly evident that trial designs offering human subjects an intervention that falls short of the best-proven approach can pass moral muster without relying on moral relativism (which would underlie an objectionable double standard), provided that certain rigorous conditions are met.

First, any application of clinical equipoise must carefully and scrupulously take into account local context. This does not, however, mean that the appropriate measure of the standard of care for determining ethical study design should be the level of access currently afforded to the study population. The so-called "local de facto" standard permits the deployment of placebos in control groups whenever the local "standard of care" is in fact no care at all.[210] But to call "no treatment" due to impoverished health care budgets the local "standard of care" is to distort the meaning of standard of care as a guiding medical norm.[211]

Recognizing that meaningful debate on this issue continues, the Commission finds that the optimal standard lies between the "best-proven" and "local de facto" interpretations: for example, a standard of care that would or should be optimal for a certain population; given their health needs and the level of available medical and logistical infrastructure, cultural practices, genetics, and economic capacity to sustain treatment into the future.

The Commission recognizes too that a rigid insistence upon a best-proven standard could have the unintended consequence of precluding meaningful research and achievable health reforms in relatively poor host countries. As Zulfiquer Bhutta, a well-known pediatrics expert, has noted, major progress in treating newborns with suspected sepsis has recently been achieved in India, but such progress would have been unlikely had the best-proven standard of care, intravenous antibiotics, been ethically required of the researchers. Instead, the researchers were able to compare an intervention against the *local* standard of care and conclude that the intervention was both accessible to the population and more effective than the local standard.[212]

FURTHER ANALYSIS AND RECOMMENDATIONS

(2) Methodological Constraints

Second, the scientific design of a trial must be adequate to yield usable results. Absent this, results will not be credible and human subjects would have been exposed to risk unnecessarily. However, there is disagreement about the level of confidence required for sound science and adequate trial design. Some insist that a placebo-controlled, double-blinded, randomized trial is the gold standard for testing new interventions or treatments. Others regard insistence upon such a design as counter-productive, and an inappropriate dismissal of potential trial results that may be somewhat less conclusive but obtainable without the potentially unacceptable ethical costs of a placebo-controlled trial.[213]

The Commission does not need to take sides in this debate to agree that the results of clinical trials must be amenable to definitive interpretation. As Robert Temple and Susan Ellenberg, clinical trial experts from the FDA, have pointed out, it is highly desirable that the meaning and significance of studies be contained within the study itself.[214] Ideally, one should not have to rely on results outside of a trial to interpret a single trial's results. In placebo-controlled trials, this additional outside evidence is not required because the study itself can show that the intervention being tested is better than no intervention. But in some designs, for example "equivalency" designs (trying to show that the intervention being studied is at least as good as an intervention already being employed), previous studies demonstrating the effectiveness of the intervention already being employed in a similar population are required. Temple and Ellenberg describe this as the problem of "assay sensitivity." Absent a placebo control or independent evidence taken from outside the trial that the intervention already being employed is effective, researchers will be unable to interpret the study's results.[215] This sort of design is appealing, however, because it does not subject trial subjects to the medical risks of a placebo control. It is, however, logistically and economically more challenging because it requires the recruitment of more subjects for trials of longer duration and greater expense.[216]

Similar worries concern equivalency trials wherein the new drug is *expected* to perform less well than the established standard, as in the perinatal AZT trials, but there are compelling policy reasons to find an alternative to the standard. In the perinatal AZT trials, the short course AZT was expected to be somewhat less effective than the "076 protocol" (the best-proven standard at the time), but also much less expensive and therefore more affordable in developing countries.[217] Indeed, the goal of the short course trials was to find an affordable and implementable regimen to prevent perinatal HIV transmission in developing countries, not to find the best such intervention worldwide. In such a context, designing a trial that would compare the established 076 protocol against short course regimens without a placebo control would have suffered from the same problem of assay sensitivity described above. If the new drug regimen had shown to be somewhat less effective than the best-proven standard, what would the researchers have learned? The intuitive answer is that the researchers would have learned that the new drug protocol was comparable to the best-proven standard and thus possibly a good candidate for public health funding; but without a comparator to judge *how* effective, decisions about adopting the intervention would be uninformed.

(3) Risk Minimization

Third, although study designs that do not use the best-proven therapy for a given condition in a given population can sometimes be justified for methodological reasons, such designs nonetheless can place subjects at risk of harm. An additional ethical constraint requires researchers to minimize the amount of harm to which subjects might be exposed.

Subjects receiving less than the optimal standard of care should not be subjected to substantially increased risk of mortality, serious morbidity, or severe discomfort. An example of a problematic study falling into the latter category was a placebo-controlled study of the anti-emetic drug, Ondansetron, which was being introduced as a remedy for nausea and vomiting associated with cancer chemotherapy.[218] At the time of the study, there were a number of approved anti-emetic drugs that were effective and had been shown in previous trials to be superior to placebo, yet subjects in the placebo arm of the Ondansetron study experienced extreme vomiting following their chemotherapy. Such a study was arguably unethical because subjects in the control arm experienced serious side effects, even if it had methodological advantages over an active control design comparing Ondansetron against another approved anti-emetic.

FURTHER ANALYSIS AND RECOMMENDATIONS III

Trials of psychiatric drugs in which some subjects receive less than the standard of care provide examples of increased risks of mortality and serious morbidity. Because depressed patients may be at risk for suicide, denying them effective treatment within a control group could place them at increased risk of death. Likewise, patients who are abruptly weaned from an anti-psychotic regimen during the washout phase of a new anti-psychotic trial might be placed at increased risk of new and possibly exacerbated psychotic episodes.

IRBs should scrutinize the elements of study design as a distinct focus of their review. Study designs that pose risks to subjects by departing from the current standard of care in control groups should be explicitly justified in the research protocol, and the anticipated additional risks to subjects should be proportional to the expected degree of individual or social benefit. Subjects who might be exposed to additional risks should be carefully examined for special vulnerabilities before trials begin, and their clinical course should be carefully monitored throughout the duration of the trial so that they can be taken off study, if necessary, to address emerging threats to their lives, health, or comfort.

The Commission finds, then, that some research designs in which control arm subjects receive less than the best-proven treatment can be ethically justified if the above criteria are all applied and clearly met. Importantly, contrary to some of the critics of the short-course perinatal HIV trials, the Commission has concluded that good ethical reasons can be found for permitting some of these trials and that one need not resort to any intellectually or morally questionable theory of ethical relativism in order to do so.

Recommendation 12: Ensure Ethical Study Design for Control Trials

When assessing how to reconcile the requirements of rigorous study design with the interests of research subjects, a nuanced approach is recommended that permits subjects to receive a placebo or an active agent that otherwise might not represent the "best-proven" approach when the site selected is ethically justifiable and the following conditions are met: a) the "best-proven" intervention is not known to be the best for a particular population due to local infrastructural, behavioral, genetic, or other relevant circumstances; and b) the scientific rationale *and* the ethical justification for the study design have undergone careful review to ensure all of

the following: i) use of placebo or other comparators is of limited duration; ii) subjects are carefully monitored; iii) rescue measures are in place should serious symptoms develop; and iv) there are established withdrawal criteria in place for subjects who experience adverse events.

8. Promoting Current Reform Efforts

Federal efforts to reform the current human subjects research protections system are already underway. In July 2011, the HHS Secretary and the OSTP Director published an ANPRM entitled *Human Subjects Research Protections: Enhancing Protections for Research Subjects and Reducing Burden, Delay, and Ambiguity for Investigators*, soliciting comments on possible revisions to the Common Rule and other provisions of the current system of protection for human research. Specifically, the ANPRM asks for public comment on "how current regulations for protecting human subjects who participate in research might be modernized and revised to be more effective."[219]

The ANPRM includes many thoughtful proposed revisions to human subjects protections. The Commission generally supports the objectives of the ANPRM and the goal of better protecting human subjects while reducing burden, delay, and ambiguity for investigators.[220] Several sections in the proposed reforms specifically are of interest to the Commission, and several topics are of particular relevance to the Commission's charge from the President. Moreover several of the recommendations made above, such as the recommendation to develop specific regulatory directions for investigators, could be included as part of the current reform effort and the Commission urges the government to consider doing so.

Ensuring Risk-Based Protections

Currently under the Common Rule, review of human research studies is undertaken by a convened IRB for all studies involving more than minimal risk to subjects or more than minor changes from previously reviewed work.[221] Studies involving no more than minimal risk or minor changes in previously approved research may be reviewed by a single IRB member through "expedited review" if the category of research appears on a published list of activities eligible for expedited review.[222] Additionally, six categories of studies are "exempt" from IRB review altogether.[223] OHRP guidance

FURTHER ANALYSIS AND RECOMMENDATIONS   III

> "The current regulations governing human subjects research were developed years ago when research was predominantly conducted at universities, colleges, and medical institutions, and each study generally took place at only a single site. Although the regulations have been amended over the years, they have not kept pace with the evolving human research enterprise, the proliferation of multi-site clinical trials and observational studies, the expansion of health services research, research in the social and behavioral sciences, and research involving databases, the Internet, and biological specimen repositories, and the use of advanced technologies, such as genomics. Revisions to the current human subjects regulations are being considered because OSTP and HHS believe these changes would strengthen protections for research subjects."
>
> Subjects Research Protections: Enhancing Protections for Research Subjects and Reducing Burden, Delay, and Ambiguity for Investigators, 76 Fed. Reg. 44,512 (July 26, 2011).

recommends—but does not require—that review of exempt studies be done by someone other than the investigator in order to confirm that the study has been appropriately classified as exempt.[224]

The current system has been criticized as "not adequately calibrating the review process to the risk of research."[225] Furthermore, many IRBs review minimal risk studies at convened meetings out of concerns for regulatory and institutional liability, rather than through expedited review. Accordingly, many social and behavioral scientists have argued that their research is over-regulated in the current system.[226]

In its first report, *New Directions: The Ethics of Synthetic Biology and Emerging Technologies*, the Commission elucidated the principle of intellectual freedom and responsibility, and its corollary principle of regulatory parsimony. These principles reflect the fact that scientific discoveries and advancements depend on the intellectual freedom of researchers coupled with the responsibility of individuals and institutions to use their creative potential in morally responsible ways. Regulatory oversight must be limited to that which is truly necessary to ensure justice, fairness, security, and safety while pursuing the public good. Consequently, the Commission supports reforming research review to appropriately calibrate the evaluation with the level of risk to human subjects. This reform also would accomplish the vital goal of allowing IRBs and institutions to focus their efforts and limited resources on studies involving higher risk. The Commission endorses the ANPRM in its goal of eliminating continuing review for certain lower-risk studies and to regularly update the list of research categories or activities that may undergo expedited review.

Also elucidated in *New Directions*, the principle of public beneficence encourages maximizing public benefits and minimizing public harm, and the principle of responsible stewardship reflects a shared obligation among members of the domestic and global communities to act in ways that demonstrate concern for those who are not in a position to represent their own interests. Both of these principles applied to human subjects research demand a thorough and more complete review of studies deemed to be greater than minimal risk to subjects. These principles also encourage a lesser regulatory burden on investigators conducting minimal risk research so that important work is not delayed through unnecessary regulatory processes and so that IRBs can focus their resources on reviewing higher risk protocols.

For exempt research, expanded and called "excused" in the ANPRM, the Commission agrees with the proposal to adopt a registration system as a reasonable strategy to track this research, provided that institutions responsible for this research retain the flexibility to require a higher standard of independent review of exempt status. Just as with determinations concerning independent review of minimal risk research, which the Common Rule currently allows to be made by expedited review,[227] decisions about exemption should be made without full IRB review so long as they are not limited solely to the discretion of the individual investigator. Data from projects, such as the "Exempt Wizard" currently being studied through the Federal Demonstration Partnership, will be useful in evaluating how registration protects subjects of exempt research.[228]

Streamlining IRB Review of Multi-Site Studies

While the current Common Rule regulations do not require IRB review by each institution involved in a multi-site study, in many—if not most—cases individual institutions[229] and IRBs independently review research protocol and informed consent documents.[230] This sometimes results in hundreds of individual reviews and, if the protocol is revised, multiple re-reviews at each site. This process is time consuming, logistically burdensome, and often duplicative for both the IRBs as well as the investigators, and there is little evidence that multiple reviews improve human subjects protections. In addition, IRBs do not always reach the same conclusions on important issues.

The Commission supports reducing time consuming duplicative IRB review where duplication serves no purpose to improve protections. Reducing unnecessary burdens to research is encouraged by the principles of intellectual freedom and responsibility and regulatory parsimony. Multiple reviews can unnecessarily delay important research, result in costly and problematic alterations to protocols and study designs, cause burdensome requirements to clear minor changes, and in some cases even jeopardize the integrity of the science. In addition, reducing burdens and barriers to effective IRB review helps speed research discoveries from the bench to the bedside and thereby effectuates the principle of public beneficence by improving public health and overall societal well-being. Although the ideal of a single IRB of record for such studies should be pursued as the ANPRM describes, the Commission believes it should not be mandated. In some situations, institutions and investigators have unresolved concerns about liability, regulatory enforcement, and implementation of a single Institutional Review Board of record, which should not be ignored.

Improving Informed Consent

The Common Rule requires legally effective informed consent from all research subjects in studies subject to the Rule, unless that requirement is waived by an IRB in accordance with pre-specified regulatory criteria.[231] However, the ANPRM notes that "consent forms may frequently fail to include some of the most important pieces of information that a person would need in order to make an 'enlightened decision' (to quote the *Nuremberg Code*) to enroll in a research study. Rather than presenting the information in a way that is most helpful to prospective subjects—such as explaining why someone might want to choose not to enroll—the forms often function as sales documents, instead of as genuine aids to good decision-making."[232]

The principle of responsible stewardship requires citizens and their representatives to think and act collectively for the common good. The government, in collaboration with institutions and investigators, should refocus on the importance of the process of informed consent lest the procedures, ostensibly derived from ethical principles, serve to obscure the values they were intended to implement. If both the government and institutions take a step back and concentrate on that which ensures actual informed consent, they may start to unravel the impediments to the process overall. That said, much like streamlining IRB review, regulating informed consent procedures must be done with great care and should be informed by data concerning what is and is not effective in achieving such consent. The Commission encourages the development of flexible guidelines and procedures for obtaining informed consent, especially in social and behavioral sciences as well as in epidemiology, public health, and quality improvement research since the issues in these kinds of research are often different than those in more classic pharmaceutical trials.

Clarifying and Harmonizing Regulatory Requirements and Agency Guidance

Although the Common Rule is widely seen as a single, comprehensive standard for protecting human subjects in research across the federal government, departments and agencies vary in their implementation of it[233] and some funders of research have not adopted it at all.[234] The ANPRM acknowledges the importance of improving the consistency of guidance on the protections of human subjects across federal departments and agencies.[235]

The Panel recommended to the Commission in its *Research Across Borders* report that "continued efforts to harmonize and guide interpretation of rules should be made a priority over creating new rules."[236] The Panel acknowledged that new rules may be needed to harmonize existing U.S. rules, but any additions should be clear, sound, and streamlined. Furthermore, harmonization would add clarity to the U.S. oversight process, particularly for international clinical trials.

The principle of responsible stewardship espouses the clarity, coordination, and accountability of regulations across the government called for both by the ANRPM and the Panel, and reinforces the compelling case for consistency, to the extent feasible, among all federal departments and agencies supporting human subjects research.

Data Collection to Enhance Adverse Event Reporting

The ANPRM also discusses the standardized collection of safety data, specifically adverse events, and suggests changes to improve the real-time, prompt collection of such data.[237] These proposed changes include using a streamlined

FURTHER ANALYSIS AND RECOMMENDATIONS III

set of data elements to satisfy most agency reporting requirements and implementing a prototype of a web-based, federal-wide portal, already in use by several agencies, to which institutions can submit safety data.[238]

This concept of centralized human subjects research data collection is also reflected in one of the Panel's recommendations to the Commission, further discussed in *Improving Accountability* above, enumerating that "[t]o enhance transparency and accountability, governments should consider requiring all greater than minimal risk research to be registered and results reported."[239]

The Commission recognizes that different agencies have different systems, but efforts should be made to report some common data elements in a unified system.

In sum, the ANPRM is an encouraging and important step forward in ensuring that human subjects in federally sponsored research are protected from harm and unethical treatment. Many of the ethical principles the Commission has previously articulated find reflection in the current proposals. The Commission expressly endorses several of the specific proposals contained in the ANPRM.

Recommendation 13: Promoting Current Federal Reform Efforts

The Commission supports the federal government's proposed reforms to:

a) Restructure research oversight to appropriately calibrate the level and intensity of the review activities with the level of risk to human subjects;

b) Eliminate continuing review for certain lower-risk studies and regularly update the list of research categories that may undergo expedited review;

c) Reduce unnecessary, duplicative, or redundant institutional review board review in multi-site studies. Regardless of the process used to review and approve studies, institutions should retain responsibility for ensuring that human subjects are protected at their location as protection of human subjects includes much more than institutional review board review. The use of a single institutional review board of record should be made the regulatory default unless institutions or investigators have sufficient justification to act otherwise;

d) Make available standardized consent form templates with clear language understandable to subjects;

e) Harmonize the Common Rule and existing regulations of the Food and Drug Administration, and require that all federal agencies conducting human subjects research adopt human subjects regulations that are consistent with the ethical requirements of the Common Rule; and

f) Work toward developing an interoperable or compatible data collection system for adverse event reporting across the federal government.

9. Following Up

The Commission recognizes that many of these recommendations have been made previously by presidentially appointed bioethics commissions and other duly appointed government advisory bodies over the past two decades, yet it found no clear response by the federal government to many of these recommendations. A primary example concerns compensation for research-related injuries, which was recommended by President Clinton's Advisory Committee on Human Radiation Experiments in 1995. Writing only six years later, in *Ethical and Policy Issues in International Research: Clinical Trials in Developing Countries*, NBAC again called for the "adequate care and compensation to participants for injuries directly sustained during research." Other commissions have more modestly recommended an investigation of the need for compensation of human subjects for injury in research trials.[240]

Recommendation 14: Responding to Recommendations

The Commission recommends that the Office of Science and Technology Policy or another appropriate entity or entities within the government respond with changes to the status quo or, if no changes are proposed, reasons for maintaining the status quo with regard to the recommendations below. Possible departments or agencies to lead the efforts include the Department of Health and Human Services, the Office for Human Research Protections, and the National Institutes of Health, as well as other funders and regulators.

FURTHER ANALYSIS AND RECOMMENDATIONS III

Table 3.3 Recommendation Follow-Up Summary

RECOMMENDATION NUMBER†	SUMMARY	OFFICE
1	Increase accountability through online access to basic human subjects research data.	OHRP/all departments and agencies that support human subjects research
2	Support the development of systematic approaches to assess the effectiveness of human subjects protections and expand support for research related to ethical and social consideration of human subjects protection.	OHRP/all departments and agencies that support human subjects research
3	Study research-related injuries to determine if there is a need for a national system of compensation or treatment for research-related injuries because subjects harmed in the course of human research should not individually bear the costs of care required to treat harms resulting directly from that research.	OSTP/HHS
4	Publicly release reasons for changing or maintaining the status quo regarding compensation or treatment for research-related injuries.	OSTP/HHS
5	Explicate the ethical underpinnings for human subjects protection requirements.	HHS/OSTP
6	Add responsibilities of investigators to the Common Rule.	HHS/OSTP
8	Adopt or revise the 2003 Department of Health and Human Services Equivalent Protections Working Group's analysis and develop a process for evaluating requests from foreign governments and other non-U.S. institutions for determinations of equivalent protections.	OHRP
9	Support further evaluation of the UNAIDS/AVAC Guidelines to provide a standardized framework for community engagement practices across research fields.	OHRP
11	Support research to develop and evaluate justifications and operational criteria for ethical site selection.	OHRP/all departments and agencies that support human subjects research
13	Develop proposed regulations to reform the current Common Rule.	OSTP/OHRP
14	Follow up.	OSTP/other appropriate entity

† Listed here are recommendations directed to the federal government only.

A response need not be unduly lengthy or provided by a single department, agency, or division. The Commission knows that analyzing and implementing (or not) the recommendations contained herein may involve work with multiple agencies across the government, and may also involve public engagement and legal processes, such as rulemaking or statutory change.

The Commission is asking that the public be informed whether the federal government intends to move forward, and if so in what way, with any or all of these recommendations.

The Commission's point, simply put, is to assist all members of society to understand the response of those entrusted with policymaking on its behalf to these specific recommendations. Repeatedly during its review, Commission members, guest speakers, and members of the public asked "what will the next 'Guatemala' be?" For many, this question related to concern about what future generations will see as "ethically impossible" in our own practices. While it is difficult to know what future generations will think, it is clear that they will be aided in their assessment by clear articulation of what this generation, through its authorized policymakers, thinks and, most importantly, how it acts.

ENDNOTES

1. Exemptions to the individual informed consent requirements permit surrogate consent of children and adults who cannot consent, automation of processes in emergency settings, and notification to or exemption from informed consent requirements for very low risk research.

2. Memorandum from President Barack Obama to Dr. Amy Gutmann, Chair, Presidential Commission for the Study of Bioethical Issues (PCSBI). (November 24, 2010). Retrieved from http://bioethics.gov/cms/sites/default/files/news/Human-Subjects-Protection-Letter-from-President-Obama-11.24.10.pdf.

3. PCSBI. (2011, September). *"Ethically Impossible" STD Research in Guatemala from 1946-1948*. Washington, DC: PCSBI.

4. The President expressly charged the Commission to engage with international experts. Memorandum from President Barack Obama, op cit. Retrieved from http://bioethics.gov/cms/sites/default/files/news/Human-Subjects-Protection-Letter-from-President-Obama-11.24.10.pdf. ("In fulfilling this charge, the Commission should seek the insights and perspective of international experts . . . .").

5. In a world of unlimited resources one could devote more funds and personnel to create myriad systemic checks on research with human subjects. Abuses or adverse events tied to human error, no matter how infrequent they may be, would surely be reduced, and those tied to design could also diminish.

6. See also The International Research Panel of the PCSBI (2011, September). *Research Across Borders: Proceedings of the International Research Panel of the Presidential Commission for the Study of Bioethical Issues*. Washington, DC: PCSBI, p. 4.

7. Advisory Committee on Human Radiation Experiments (ACHRE). (1995). *Final Report of the Advisory Committee on Human Radiation Experiments*. New York: Oxford University Press; National Bioethics Advisory Commission (NBAC). (2001). *Ethical and Policy Issues in International Research: Clinical Trials in Developing Countries*. Bethesda, MD: NBAC; National Commission for the Protection of Human Subjects of Biomedical and Behavioral Research. (1976). *Research Involving Prisoners* (DHEW Publication No. (OS) 46-131). Washington, DC: Department of Health, Education, and Welfare (DHEW); President's Commission for the Study of Ethical Problems in Medicine and Biomedical and Behavioral Research. (1982). *Compensating for Research Injuries*. Washington, DC: DHEW.

8. Equivalent Protections Working Group. (2003). *Report of the Equivalent Protections Working Group*. Washington, DC: Department of Health and Human Services (HHS); NBAC. (2001). *Ethical and Policy Issues in International Research: Clinical Trials in Developing Countries*. Bethesda, MD: National Bioethics Advisory Commission.

9. NBAC, *Ethical and Policy Issues in International Research: Clinical Trials in Developing Countries*, op. cit; NBAC. (2001). *Ethical and Policy Issues in Research Involving Human Participants*. Bethesda, MD: NBAC.

10. Jonas, H. (1969). Philosophical reflections on experimenting with human subjects. *Daedalus: Journal of the American Academy of Arts and Sciences*, 98, 245.

11. Franklin, B., and J. Sparks. (1839). *The Works of Benjamin Franklin: Vol. VIII*. Boston, MA: Hilliard, Gray, and Company, p. 418.

ENDNOTES

[12] Franklin, of course, employs the term "moral science" in the usage of the day, namely, as the science of morality. For the Commission, the term is synonymous with ethically sound science. In the spirit of the Enlightenment, philosophers like Frances Hutchison and Jeremy Bentham, so influential in the thinking of the American Founders, wanted to find an empirical or "naturalistic" grounding for morality. A still more provocative matter, to which Franklin's musing alludes only indirectly, is the question whether progress in "Experimental Researches into Nature" might also somehow make experimental science more ethically sound. It is a question that has preoccupied Americans ever since and the answer has vacillated. For example, by the middle and late 19th century the idea that experimental science and moral science co-evolved was prevalent among public intellectuals, including academics and clergy. But confidence that that is the case seems to have waned in our own time. For further discussion of these principles, see, e.g., Emanuel, E.J., Wendler, D., and C. Grady. (2000). What makes clinical research ethical? *JAMA*, 283, 2701-2711; PCSBI. (2011, September); *"Ethically Impossible" STD Research in Guatemala from 1946-1948*. Washington, DC: PCSBI (describing long-standing ethical traditions for medical research); The National Commission for the Protection of Human Subjects of Biomedical and Behavioral Research. (1979). *The Belmont Report: Ethical Principles and Guidelines for the Protection of Human Subjects of Research* (DHEW Publication OS 78-0012). Washington, DC: Department of Health, Education, and Welfare, pp. 4-6 (describing principles of autonomy, beneficence and justice); World Medical Association. (2008). *Code of Ethics of the World Medical Association: Declaration of Helsinki*. Ferney-Voltaire, France: World Medical Association.

[13] Franklin made strenuous attempts to contribute to medical science. He engaged in electrotherapy experiments with persons suffering from paralysis, at their request. Unfortunately, as he himself noted, his efforts failed to produce a cure. Franklin, B. (1769). *Experiments and Observations on Electricity made at Philadelphia in America* (4th ed.). London: David Henry, pp. 359-361. Anticipating the importance of data for validating public health interventions, Franklin also compiled statistics on the effectiveness of smallpox vaccination. Franklin's belief in reason and public demonstration as a corrective to dogmatism led him to participate in a French commission that engaged in experiments to disprove the "animal magnetism" claims of Franz Anton Mesmer's medical cult. Whether Franklin obtained the consent of the experimental subjects, which included members of his own household, is unknown. Gensel, L. (2005). The medical world of Benjamin Franklin. *Journal of the Royal Society of Medicine*, 98(1), 534-538.

[14] Normally, informed consent must be obtained prospectively from each subject. In some research, like that involving anonymous information or information that is de-identified under the HIPAA Privacy Rule (45 C.F.R. §164.512), or that involving minimal risk, certain routine surveys and other low-risk categories, current regulations do not require informed consent. See 45 C.F.R. §§ 46.101(b), 46.116(c)-(d). Generally, these exclusions have not been controversial and, for many, they reflect society's expectations. A few poignant examples demonstrate how complicated decisions about informed consent can be. For example, in testimony to the Commission, Carletta Tilousi, a member of the Havasupai tribe from the southwestern United States, described the pain and suffering she and others in her tribe experienced when informed consent was not sought from them for secondary research use of data and samples. Tilousi, C., Member of the Havasupai Tribe. (2011). Community Engagement—Needs, Models and U.S. Actions. Presentation to the PCSBI, August 30,

2011. Retrieved from http://bioethics.gov/cms/node/319 (accessed December 5, 2011). Under current law, "what constitutes adequate informed consent is unsettled. Federal regulations require informed consent when identifiable biospecimens are collected for research purposes, but such regulations provide little guidance on how to obtain informed consent for future, unspecified uses." See Mello, M.M., and L.E. Wolf. (2010). The Havasupai Indian tribe case—Lessons for research involving stored biologic samples. *New England Journal of Medicine*, 363(3), 205.

[15] Lemonick, M.D., and A. Goldstein. (2002, April 22). At your own risk. *TIME*. Retrieved from http://www.time.com/time/magazine/article/0,9171,1002263,00.html (accessed December 2, 2011) (describing the 2001 death of a healthy woman following participation in an early-stage asthma treatment study); Suntharalingam, G., et al. (2006). Cytokine storm in a phase 1 trial of the anti-CD28 monoclonal antibody TGN1412. *The New England Journal of Medicine*, 355, 1018-1028. (describing a 2006 Phase 1 drug trial that hospitalized six otherwise healthy men with multiple organ system failure due to unknown properties of the chemical agent).

[16] Memorandum from President Barack Obama to Dr. Amy Gutmann, Chair, PCSBI. (2010, November 24). Retrieved from http://bioethics.gov/cms/sites/default/files/news/Human-Subjects-Protection-Letter-from-President-Obama-11.24.10.pdf.

[17] Read-out of the President's Call with Guatemalan President Colom. (2010, October 1). Retrieved from http://www.whitehouse.gov/the-press-office/2010/10/01/read-out-presidents-call-with-guatemalan-president-colom.

[18] PCSBI, *"Ethically Impossible,"* op cit.

[19] Memorandum from President Barack Obama to Dr. Amy Gutmann, op cit.

[20] This policy required "voluntary agreement based on informed understanding" from all research subjects, written consent when a physician deemed the research "unusually hazardous," and, beginning in 1954, written consent was required from all "healthy, 'normal' subjects of research." ACHRE, op cit.; it is not, however, the earliest known federal policy requiring informed consent from subjects. See PCSBI, *"Ethically Impossible,"* op cit. (discussing policy decision requiring informed consent in prison research in 1945); ACHRE, op cit. (discussing the evolution of policies across the federal government for consent and other human subjects protections in the 1940s through the 1970s).

[21] See Drug Amendments of 1962, Public Law No. 87-781 (1962), *codified at* 21 U.S.C. 355 (requiring informed consent in certain FDA-regulated research); 21 C.F.R. 50 (same); National Research Act of 1974, Public Law No. 93-348 (1974), *codified at* 42 U.S.C. 218 (requiring informed consent and independent review in HHS-sponsored research). See also ACHRE. (1995), op cit., pp. 64-67, 83-129, 171-195.

[22] Coleman, C., et al. (2005). *The Ethics and Regulation of Research with Human Subjects*. Newark, NJ: LexisNexis. Before the National Commission, the Department of Health, Education and Welfare convened the Tuskegee Syphilis Study Ad Hoc Advisory Panel in 1972. This Panel found that the men involved in the Tuskegee study had not been informed of the actual purpose of the study, had been mislead, and thus had been unable to provide informed consent. The Panel concluded that the study was "ethically unjustified" and recommended stopping the study immediately. A month later, the study was terminated.

CDC. (2011, June 15). *U.S. Public Health Service Syphilis Study at Tuskegee.* Retrieved from http://www.cdc.gov/tuskegee/timeline.htm (accessed December 5, 2011).

[23] Citro, C.F., Ilgen, D.R., and C.B. Marrett, (eds.). (2003). *Protecting Participants and Facilitating Social and Behavioral Sciences Research.* Washington, DC: The National Academies Press, pp. 45-53.

[24] Glickman, S.W., et al. (2009). Ethical and scientific implications of the globalization of clinical research. *The New England Journal of Medicine,* 360(8), 816-823.

[25] Office of the Inspector General (OIG). (2010, June). *Challenges to FDA's Ability to Monitor and Inspect Foreign Clinical Trials* (OEI-01-08-00510). Retrieved from http://oig.hhs.gov/oei/reports/oei-01-08-00510.pdf.

[26] PCSBI, *Research Across Borders,* op cit., p. 2.

[27] PCSBI, *Research Across Borders,* op cit. Public comments on the report were solicited through a notice in the Federal Register, at Commission meetings, and through the Commission's website. Request for Comments on Research Without Borders: Proceedings of the International Research Panel of the Presidential Commission for the Study of Bioethical Issues, 76 Fed. Reg. 55,914 (Sept. 9, 2011).

[28] "[Rules] may be interpreted or implemented differently as a result of complex cultural, political, and economic influences." Ibid, pp. 5-6.

[29] Ibid, pp. 8-11.

[30] See www.clinicaltrials.gov for more information. See also the section on *Improving Accountability.*

[31] Ibid, p. 11.

[32] Ibid, pp. 8-11. For example, as discussed in recent regulatory reform efforts, multiple IRBs for multi-site studies might actually be leading to weaker protections for subjects than fewer reviews with greater responsibility on the part of the IRBs involved. Subjects Research Protections: Enhancing Protections for Research Subjects and Reducing Burden, Delay, and Ambiguity for Investigators, 76 Fed. Reg. 44,513 (July 26, 2011); Boscheck, R., et al. (2008). *Strategies, Markets and Governance: Exploring Commercial and Regulatory Agendas.* Cambridge: Cambridge University Press, p. 285 (using the term "process fatigue" to reflect the burden of the proliferation of monitoring mechanisms associated with EU constitutional governance).

[33] PCSBI, *Research Across Borders,* op cit., p. 4.

[34] 45 C.F.R. Parts 46, Parts A through D, 160, and 164 and, through the Food and Drug Administration (FDA) at 21 C.F.R. Parts 50 and 56. Other federal departments and agencies joined HHS in adopting a uniform set of rules for the protection of human subjects, the "Common Rule," identical to subpart A of 45 C.F.R. part 46 of the HHS regulations. See more extensive discussion of the current regulations in the section on *Promoting Current Reform Efforts.*

[35] Subjects Research Protections: Enhancing Protections for Research Subjects and Reducing Burden, Delay, and Ambiguity for Investigators, 76 Fed. Reg. 44,512-44,530 (July 26, 2011).

[36] Ibid., p. 44,512.

37 Request for Comments on Human Subjects Protections in Scientific Studies, 76 Fed. Reg. 11,482 (March 2, 2011).

38 Bartlett, E. E. (2008). International analysis of institutional review boards registered with the Office for Human Research Protections. *Journal of Empirical Research on Human Research Ethics*, 3(4), 49-56; Bhutta, Z. A. (2002). Ethics in international health research: A perspective from the developing world. *Bulletin of the World Health Organization*,114-120; Glickman, S. W., et al. (2009). Ethical and scientific implications of the globalization of clinical research. *The New England Journal of Medicine*, 360(8), 816-823; Lansang, M. A., and F.P. Crawley. (2000). The ethics of international biomedical research. *British Medical Journal*, 321(7264), 777-778; McCarthy, J., Wilenzick, M., and C. Cuenot. (2010). The globalization of clinical trials – Challenges, opportunities and a path forward. *International In-house Counsel Journal*, 3(11), 1-13; Office for Human Research Protections, U.S. Department of Health and Human Services. (2012). International Compilation of Human Research Standards. Retrieved from http://www.hhs.gov/ohrp/international/intlcompilation/intlcompilation.html (accessed December 7, 2011). The Commission is especially grateful to Edward E. Bartlett, PhD, International Human Research Liaison, Office for Human Research Protections and the International Working Group for providing guidance and assistance in understanding human subjects protections around the globe. The Commission reached out to the heads of the 18 federal agencies and departments that have adopted the Common Rule. They are: (1) Department of Agriculture (USDA); (2) Department of Commerce (DOC); (3) Department of Defense (DOD); (4) Department of Education (ED); (5) Department of Energy (DOE); (6) Department of Health and Human Services (HHS); (7) Department of Homeland Security (DHS); (8) Department of Housing and Urban Development (HUD); (9) Department of Justice (DOJ); (10) Department of Transportation (DOT); (11) Department of Veterans Affairs (VA); (12) Agency for International Development (USAID); (13) Consumer Product Safety Commission (CPSC); (14) Environmental Protection Agency (EPA); (15) National Aeronautics and Space Administration (NASA); (16) National Science Foundation (NSF); (17) Social Security Administration (SSA); and (18) the Central Intelligence Agency (CIA).

39 ACHRE. (1995). *Final Report of the Advisory Committee on Human Radiation Experiments*. New York, NY: Oxford University Press.

40 NBAC. (2001). *Ethical and Policy Issues in Research Involving Human Participants*. Bethesda, MD: NBAC; NBAC. (2001). *Ethical and Policy Issues in International Research: Clinical Trials in Developing Countries*. Bethesda, MD: NBAC.

41 Institute of Medicine (IOM). (2003). *Responsible Research: A Systems Approach to Protecting Research Participants*. Washington, DC: National Academies Press.

42 See generally, PCSBI. (2010, December). *New Directions: The Ethics of Synthetic Biology and Emerging Technologies*. Washington, DC: PCSBI, p. 17. (as with synthetic biology, "Ongoing assessment and review is required in several areas to avoid unnecessary limits on science"); PCSBI, *Research Across Borders*, op cit., pp. 8-11; "The stacking of regulations within and across agencies and countries increases the burden without any benefit to the subjects and in many cases results in increases costs and decreased research productivity." DeCrappeo, A.P., President, Council on Governmental Relations. (2011). Comments submitted to the PCSBI, May 2, 2011.

ENDNOTES

43 IOM, op cit., p. 9.

44 Bierer, B.E., Chair, Secretary's Advisory Committee on Human Research Protections (SACHRP). (2011). Comments submitted to PCSBI, August 5, 2011. Retrieved from http://www.hhs.gov/ohrp/sachrp/commsec/commentspcsbi.pdf.pdf (accessed December 2, 2011).

45 Memorandum from President Barack Obama to Dr. Amy Gutmann, op cit.

46 PCSBI, *"Ethically Impossible,"* op cit.

47 See, e.g., International Convention on Civil and Political Rights, *adopted and opened for signature* Dec. 16, 1996, 999 U.N.T.S. 171. Note, however, that although the ICCPR Article 7 guarantees that "[n]o one shall be subjected to torture or to cruel, inhuman or degrading treatment or punishment. In particular, no one shall be subjected without his free consent to medical or scientific experimentation," the United States has ratified the treaty, but reserved articles 1-27 as non-self-executing. Consequently, there is no private right of action, or recourse to the U.S. state or federal courts, for violating Article 7. 138 Cong. Rec. S. 4781.

48 Some U.S. states also impose one or more of these standards on all research within their borders. Schwartz, J. (2001). Oversight of Human Subject Research: The Role of the States. In H. Shapiro, et al. (Eds), *Ethical and Policy Issues in Research Involving Human Participants*. (pp. M1-M20). See also Cal. Health and Safety § 24175 (2010). (mandating that, with very narrow exceptions, no person shall be subjected to any medical experiment unless the informed consent of such person is obtained); Md. Code Ann., Health-Gen § 13-2002(a) (2011). (adopting the Common Rule as a matter of state law, stating, "A person may not conduct research using a human subject unless the person conducts the research in accordance with the federal regulations on the protection of human subjects"); Va. Code Ann. §§ 32.1-162.18, 32.1-162.19 (2011). (requiring for the performance of human subjects research (1) "informed consent must be obtained" and (2) that research and research proposals be subjected to the review of a human subject review committee). State law sometimes mandates additional protections; Common Rule regulations do not preempt state or local laws that provide additional protections for human subjects. 45 C.F.R. § 46.101(f).

49 See generally 45 C.F.R. pt. 46, Subpart A. The policy is known as the Common Rule because it was promulgated jointly in 1991 as a single set of regulations by 15 different departments and agencies. Since 1991, three additional departments and agencies have signed onto it. See Federal Policy for the Protection of Human Subjects ("Common Rule"), 56 Fed. Reg. 28003 (June 18, 1991). Institutions must certify that proposed research with human subjects, that is subject to the regulations, has been reviewed and approved by an IRB. The regulations set forth specific requirements for IRB membership, as well as criteria for IRB approval of research. IRBs are empowered to approve, require modifications of, or disapprove covered research activities, and are required to conduct continuing review of ongoing research at least annually. They also have authority to suspend or terminate approval of research not conducted in accordance with the IRB's requirements or that has been associated with unexpected serious harms to subjects. Institutions must provide written assurance to the agency sponsoring the research that they will comply with the Common Rule. They must adopt a statement of principles for protecting human subjects; designate a responsible IRB; retain information on IRB membership; establish written procedures for IRB review and reporting unanticipated problems, noncompliance, and suspension or termination of IRB approval. 45 C.F.R. pt 46, Subpart A.

50   FDA has not adopted the Common Rule *per se*, but has its own regulations for human subjects protection. Like the Common Rule FDA regulations require informed consent and independent review. The regulations also specify safety data and other details that must be provided for FDA review prior to initiation of most clinical investigations, describe the responsibilities of sponsors and investigators, set forth standards to help assure that studies will be acceptable to support marketing approval in order to avoid exposing subjects to unwarranted burden and risk, and address conflicts of interest. 21 C.F.R. pt. 50, pt. 56, pt. 312, and pt. 812.

51   See also Congressional Research Service. (2005). CRS Report for Congress: Federal Protection for Human Research Subjects- An Analysis of the Common Rule and the HIPAA Privacy Rule. June 2. Retrieved from http://www.fas.org/sgp/crs/misc/RL32909.pdf (accessed December 6, 2011); Faden, R.R. and T.L. Beauchamp. (1986). *A History and Theory of Informed Consent*. New York, NY: Oxford University Press; Porter, J.P. and G. Koski. (2008). Regulations for the Protection of Humans in Research in the United States: The Common Rule. In E.J. Emanuel, et al. (Eds). *The Oxford Textbook of Clinical Research Ethics*. New York: Oxford University Press, (pp. 156-167). State law also regulates certain research outside the reach of federal law, though not all states regulate human subjects research. Those that do, generally follow the example of federal law, with a focus on independent prior review and legally effective informed consent. For example, New York, one of the earliest adopters of relevant state law, mandates that "no human research may be conducted in this state in the absence of the voluntary informed consent subscribed to in writing by the human subject." N.Y. Public Health Law § 2442 (Consol. 2011). Notably, New York law defines "human research" as "any medical experiments, research, or scientific or psychological investigation, which utilizes human subjects," which, unlike federal law, brings both privately and publicly funded human research activities under its purview. N.Y. Public Health Law § 2441 (Consol. 2011). New York law also requires institutions that conduct or propose to conduct human research to establish a human research review committee. That committee reviews research and approves "a statement of principle and policy," issued by the institution conducting or authorizing human research, "in regard to the rights and welfare of human subjects in the conduct of human research." N.Y. Public Health Law § 2444 (Consol. 2011). See also Va. Code Ann. § 32.1-162.20 (2011); MD Code Ann., Health–Gen. §13-2002 (2011); MD Code Ann., Health–Gen. §13-2001 (2011). By contrast to New York, Virginia, and Maryland, California does not require prior review by a research ethics committee, but requires that subjects be given an "experimental subject's bill of rights" and provides civil and criminal penalties for those who negligently or willfully fail to obtain informed consent. Cal Health & Safety § 24170-24179.5 (2010). Additionally, some states have also chosen to regulate in particular areas—including human cloning, genetic research, and stem cells—or to provide additional protections for certain classes of research participants—including research with prisoners, children, patients in psychiatric facilities, individuals in facilities for the developmentally disabled, or fetuses. See, e.g., Va. Code Ann. §§ 32.1-162.22 (2011).

52   See e.g., Aristotle. (1999). *Nicomachean Ethics*. T. Irwin, (Ed.). Indianapolis, IN: Hackett Publishing; Kant, E. (1998). *Groundwork of the Metaphysics of Morals*. Gregory, M. (Ed.). Cambridge: Cambridge University Press; Pellegrino, E.D. and D.C. Thomasma. (1993). *The Virtues in Medical Practice*. New York, NY: Oxford University Press, pp. 3-29.

ENDNOTES

[53] See e.g., American Society for Investigative Pathology (ASIP). (2011). ASIP By Laws. February Retrieved from http://www.asip.org/gov/documents/bylaws.PDF; Glaxo-Smith Klein. (2010). GSK Corporate Responsibility Report. Retrieved from http://www.gsk.com/responsibility/downloads/GSK-CR-2010-Report.pdf; Yale University. (2010, May 1). Policy 1360: Human Research Protection. Retrieved from http://www.yale.edu/policy/1360/1360.pdf.

[54] See e.g., AstraZeneca. *Code of Conduct.* Retrieved from http://www.astrazeneca.com/Responsibility/Code-policies-standards; PhRMA. (2011, July). *PhRMA Principles on Conduct of Clinical Trials: Communication of Clinical Trial Results.* Retrieved from http://www.phrma.org/sites/default/files/105/042009_clinical_trial_principles_final.pdf.

[55] Bayer. (2011, January 20). *Bayer: Science for a Better Life Mission and Values.* Retrieved from http://www.bayer.com/en/mission---values.aspx; Bristol-Myers Squibb. (2010). *Bristol-Myers Squibb Standards of Business Conduct and Ethics.* Retrieved from http://www.bms.com/Documents/ourcompany/SBCE_2010.pdf; Novartis. (2001, June). *Novartis Code of Conduct.* Retrieved from http://www.corporatecitizenship.novartis.com/downloads/business-conduct/code_of_conduct.pdf; PhRMA, op cit.; Sanofi-Aventis. (2011, August 19). *Sanofi US Code of Business Conduct.* Retrieved from http://www.sanofi.us/l/us/en/layout.jsp?scat=DD8ABF97-8898-4629-8CBF-931915609029 (Note: actual document not publicly available.).

[56] AstraZeneca, op cit., (stating that the key issues in the code address the fundamental principle that members must "conduct clinical research in a manner that recognizes the importance of protecting the safety of and respecting research participants").

[57] Responsibility of Applicants for Promoting Objectivity in Research for which Public Health Service Funding is Sought and Responsible Prospective Contractors. 76 Fed. Reg. 53,256 (August 25, 2011). (to be codified at 42 C.F.R. pt. 50, 45 C.F.R. pt. 94); Standards for Privacy of Individually Identifiable Health Information, 45 C.F.R. pt. 160 and pt. 164.

[58] See Appendix II (explaining data collection methods).

[59] The Human Subjects Research Landscape Project includes research that was not reported because it is classified or because of national security concerns. Classified research is generally not disclosed to the public. However, such research must undergo review by an IRB. For example, at DOD an IRB must have at least five members, one of whom must be non-governmental, and classified research may never be approved using expedited review procedures. Office of the Under Secretary of Defense. (2011). *Research Regulatory Oversight Office, Human Research Protection Program, Operating Instruction.* Retrieved from: http://home.fhpr.osd.mil/Libraries/DoDD-RROO/Signed_OI_17_June_2011.sflb.ashx. CIA and DHS are subject to similar rules aligning with the Common Rule requirements of IRB review and prohibiting expedited review of classified research. Executive Order 12333, Sec. 2.10; Letter from V. Sue Bromley, CIA, to Dr. Amy Gutmann, PCSBI. (May 16, 2011); DHS. (2007). DHS Management Directive 026-04: Protection of Human Subjects, May 25, 2007. Retrieved from www.dhs.gov/xlibrary/assets/foia/mgmt-directive-026-04-protection-of-human-subjects.pdf.

[60] See Tables I.17, I.18, and I.19 (showing low response rates from some agencies for number of subjects, exempt/non-exempt status, and number of sites). Similarly, DHS for example, did not provide funding data because "it would be overly burdensome, and in many cases not feasible, to obtain the requested award information [e.g., award ID and total award amount] for the years requested by the Commission and align the human subject research data with the DHS award information obtained through OMB [i.e., USASpending.gov])." Letter from Richard Legault, DHS, to Valerie Bonham, PCSBI. (2011, September 29). ED also could not link a significant portion of extramural funding information to study-level data, but is working to improve by developing grant and contract management systems designed "to enable coordinated implementation of [ED human subjects protection regulations] as part of the broader grant and contract management system." Jeffery Rodamar, ED, to Michelle Groman, PCSBI. (2011, September 6). E-mail Correspondence. DOD also could not readily link funding data to nearly all of their reported studies. Michelle Groman, PCSBI, to Patty Decot, DOD. (2011, October 27). Email Correspondence.

[61] Honey, K. (2011). True dedication to clinical research: The Clinical Center of the National Institutes of Health receives the 2011 Mary Woodard Lasker Award for Public Service. *Journal of Clinical Investigation,* 121(10), 3778-3781.

[62] Silver, H. et al. (2011). Social and Behavioral Science Research in the FY 2012 Budget. In *AAAS Report XXXVI: Research & Development FY 2012*. pp. 205-213. Retrieved from http://www.aaas.org/spp/rd/rdreport2012/12pch19.pdf; Kobor, P., Wurtz, S., and D. Johnson. (1999). Behavioral and Social Sciences in the FY2000 Budget. In *AAAS Report XXIV: Research & Development FY 2000*. Retrieved from http://www.aaas.org/spp/rd/xxiv/chap20.htm.

[63] For example, a violation of IRB approval of informed consent may include a lack of a statement describing the extent to which confidentiality of records identifying the subject will be maintained. See generally Office of Human Research Protections (OHRP). (2011). 2011 Determination Letters. Retrieved from http://www.hhs.gov/ohrp/compliance/letters/index.html.

[64] Borror, K., et al. (2003). A review of OHRP compliance oversight letters. *IRB: Ethics & Human Research,* 25(5), 1. (reviewing OHRP compliance letters issued between October 1, 1998 and June 20, 2002); Weil, C., et al. (2010, March-April). OHRP compliance oversight letters: An update. *IRB: Ethics & Human Research,* 32(2), 1-6 (reviewing OHRP compliance letters issued between August 1, 2002 and August 31, 2007).

[65] Porter, J.P. and G. Koski, op cit., pp. 156-167.

[66] Human Subjects Research Protections: Enhancing Protections for Research Subjects and Reducing Burden, Delay, and Ambiguity for Investigators, 76 Fed. Reg. 44,512 (July 26, 2011).

[67] Concerns such as these are one factor motivating the promulgation of international human rights treaties ensuring universal rights to bodily integrity and health. See e.g., International Convention on Civil and Political Rights, *adopted and opened for signature* Dec. 16, 1996, 999 U.N.T.S. 171; International Covenant on Economic, Social, and Cultural Rights, *adopted and opened for signature* Dec. 16, 1996, 993 U.N.T.S. 3.

ENDNOTES

68 Dieppe, P. (2004, October 14). Lessons from the withdrawal of rofecoxib. *British Medical Journal,* 329, 867-868; Saul, S. (2008, April 15). Ghostwriters Used in Vioxx Studies, Article Says. *New York Times.* Retrieved from http://www.nytimes.com/2008/04/15/business/15cnd-vioxx.html?pagewanted=all (last accessed December 6, 2011); Harris, G. (2010, February 19). Research Ties Diabetes Drug to Heart Woes. *New York Times.* Retrieved from http://www.nytimes.com/2010/02/20/health/policy/20avandia.html?pagewanted=all (last accessed December 6, 2011); Steinbrook, R. (2004). Public registration of clinical trials. *New England Journal of Medicine,* 351(4), 315-317.

69 Food and Drug Modernization Act of 1997 § 113, 42 U.S.C. § 282(i) (2011).

70 Food and Drug Administration Amendments Act of 2007 § 801, 42 U.S.C. § 282(j) (2011).

71 Me. Rev. Stat. Ann. tit. 22, § 2700-A (2011); International Federation of Pharmaceutical Manufacturers & Associations. (2005); Patients and Physicians To Gain Easy Access To Clinical Trials Information Via New Ifpma Search Portal [Press release]. Retrieved from http://clinicaltrials.ifpma.org/clinicaltrials/fileadmin/files/pdfs/EN/CTP_Release_0_EN.pdf (last accessed December 6, 2011); World Health Organization. (2005). Resolutions and Decisions: Ministerial Summit on Health Research. Retrieved from https://apps.who.int/gb/ebwha/pdf_files/WHA58/WHA58_34-en.pdf (last accessed December 6, 2011).

72 See *Research Across Borders*, op cit., p. 10. Australian New Zealand Clinical Trials Registry. (2011). Retrieved from http://www.anzctr.org.au/; Ministerio da Saude. Registro Brasileiro de Ensaios Clinicos. (n.d.). Retrieved from http://www.ensaiosclinicos.gov.br/ (accessed December 6, 2011); National Institute of Medical Statistics (Indian Council of Medical Research). (2011). Clinical Trials Registry – India. Retrieved from http://ctri.nic.in/Clinicaltrials/login.php (accessed December 6, 2011).

73 World Medical Association (WMA). (2008). WMA Declaration of Helsinki – Ethical Principles for Medical Research Involving Human Subjects. Article 19. Retrieved from http://www.wma.net/en/30publications/10policies/b3/index.html.

74 OMB Memorandum from Vivek Kundra to the Heads of Departments and Agencies. (June 1, 2009). Guidance on Data Submission under the Federal Funding Accountability and Transparency Act (FFATA). Retrieved from http://www.whitehouse.gov/sites/default/files/omb/assets/memoranda_fy2009/m09-19.pdf (last accessed December 6, 2011); Public Law No. 109-282, as amended by section 6202(a) of Public Law 110-252 (see U.S.C. 6101 note).

75 American Recovery and Reinvestment Act of 2009, Public Law No. 111-5 § 3001(c)(4), 42 U.S.C. § 300jj-11(c)(4) (2011).

76 COHRED: Council on Health Research for Development. (2011). Retrieved from http://www.cohred.org/ (accessed December 6, 2011).

77 Consolidated Appropriations Act of 2008, Division G, Title II, Public Law No. 110-161; NIH. (2008). Revised Policy on Enhancing Public Access to Archived Publications Resulting from NIH-Funded Research. Retrieved from http://grants.nih.gov/grants/guide/notice-files/NOT-OD-08-033.html (accessed November 29, 2011). ("The Director of the National Institutes of Health shall require that all investigators funded by the NIH submit or have submitted for them to the National Library of Medicine's PubMed Central an electronic version of their final, peer-reviewed manuscripts upon acceptance for publication, to be made

publicly available no later than 12 months after the official date of publication: Provided, That the NIH shall implement the public access policy in a manner consistent with copyright law."). For more information see http://publicaccess.nih.gov/policy.htm (accessed December 6, 2011).

[78] Request for Information: Public Access to Peer-Reviewed Scholarly Publications Resulting From Federally Funded Research, 76 Fed. Reg. 68,518-68,520 (Nov. 4, 2011). The government's recognition of openness and transparency as a tool for accountability is well documented. See Presidential Memorandum – Managing Government Records, (2011, November 28). Retrieved from http://www.whitehouse.gov/the-press-office/2011/11/28/presidential-memorandum-managing-government-records (directing agencies to make records more available and accessible to the public); Statement of Cass Sunstein, Administrator of the Office of Information and Regulatory Affairs of the Office of Management and Budget. (November 28, 2011). We Can't Wait: Bringing Records Management into the Twenty-first Century. Retrieved from http://www.whitehouse.gov/blog/2011/11/28/we-cant-wait-bringing-records-management-twenty-first-century-0 (describing how agency information "provide[s] a prism through which future generations will view, understand and learn from the actions of the current generation); *The Open Government Partnership—National Action Plan for the United States of America*. (2011, September 20). Retrieved from http://www.whitehouse.gov/sites/default/files/us_national_action_plan_final_2.pdf. For more information on existing disclosure requirements, see, e.g., Office of Management and Budget: Open Government website. Retrieved from http://www.whitehouse.gov/omb/open.

[79] See DOE Human Subjects Research Database (HSRD). Retrieved from http://www.orau.gov/HsrdReport/ (accessed December 6, 2011). The HSRD database includes information on all research projects involving human subjects that are not exempt by their local Institutional Review Board (IRB). The data is reported to the DOE by the project's principal investigator on an annual basis, and project data is available back to 1995.

[80] See Research Portfolio Online Reporting Tools (RePORTER), NIH. Retrieved from http://projectreporter.nih.gov/reporter.cfm (accessed December 6, 2011).

[81] The Commission recognizes that in many cases agencies may not have access to this information when research is funded through grant or contract but conducted by third parties. NIH. (2011). *NIH Grants Policy Statement*. October 1, 2011. Retrieved from http://grants.nih.gov/grants/policy/nihgps_2011/nihgps_2011.pdf. (demonstrating the arms length nature of awards, it states "[o]verall responsibility for successfully implementing an NIH grant is a shared responsibility of the [program director]/[principal investigator], the [authorized organization representative], and the research administrator. As key members of the grant team, they respectively lead the scientific and administrative aspects of the grant.).

[82] AAHRPP. (2011). AAHRPP Our Mission, Vision, and Values. Retrieved from http://www.aahrpp.org/www.aspx?PageID=5 (accessed December 6, 2011); IOM. (2003). *Exploring Challenges, Progress, and the New Models for Engaging the Public in the Clinical Research Enterprise: Clinical Research Roundtable Workshop Summary*. Washington, DC: The National Academies Press, pp. 44-45.

[83] PCSBI, *Research Across Borders*, op cit., p. 10.

ENDNOTES

[84] Public evaluation of research results is also important. The Commission recognizes the programs already in place to accomplish this. See discussion in Improving Accountability section of this report regarding NIH public access programs, clinicaltrials.gov, and grant-wide proposals to require free online access to published results of federally funded research.

[85] ClinicalTrials.gov today collects most of the elements the Commission specifies. It includes 1) title, 2) principal investigator and institutional affiliation, 3) collaborating institutions, 4) location of research activities, and 5) federal funding source. See http://www.clinicaltrials.gov/ for more information. Reliance on this existing program may be impractical for all agencies as it is oriented to biomedical research, but the Commission encourages its use to consolidate reporting requirements and simplify public access.

[86] OHRP. About OHRP. (n.d.). http://www.hhs.gov/ohrp/about/index.html (accessed December 6, 2011).

[87] For example, OHRP manages the Federalwide Assurance system. See OHRP. (2011, June 17). Federalwide Assurance for the Protection of Human Subjects. Retrieved from http://www.hhs.gov/ohrp/assurances/assurances/filasurt.html (accessed December 6, 2011).

[88] Abbott, L., and C. Grady. (2011). A systematic review of the empirical literature evaluating IRBs: What we know and what we still need to learn. *Journal of Empirical Research on Human Research Ethics*, 3-19; Emanuel, E.J. (2002). The relevance of empirical research for bioethics. In Stepke, F.L. and L.A. Corbinos (Eds), *Interfaces Between Bioethics and the Empirical Social Sciences*. Buenos Aires, Argentina: World Health Organization. Letter from Barbara E. Bierer, Chair, SACHRP, to Kathleen Sebelius, Secretary, HHS. (August 5, 2011). Retrieved from http://www.hhs.gov/ohrp/sachrp/commsec/commentspcsbi.pdf.pdf ("There is an inadequate evidence base to inform regulation and best practices in many areas of human subjects protections. Increased federal support for research to enhance this evidence base is essential to facilitate improvements in human subjects research regulations and practices.").

[89] Sugarman, J., and D. Sulmasy. (Eds) (2001). *Method in Medical Ethics*. Washington, DC: Georgetown University Press, 2nd edition.

[90] Emanuel, E.J. (2002). The relevance of empirical research for bioethics. In F.L. Stepke and L.A. Corbinos (eds.), *Interfaces Between Bioethics and the Empirical Social Sciences*. Buenos Aires: World Health Organization.

[91] Collins, F., Director, NIH. (2011). Bioethics Research at the NIH. Presentation to PCSBI, February 28, 2011. Retrieved from http://bioethics.gov/cms/node/187 (accessed December 6, 2011).

[92] The majority of the discussion that follows applies directly to medical research. The majority of physical and psychological harms that arise in human subjects research appear in this arena. Insofar as such harms arise in other fields of human subjects research, however, the Commission believes that the ethical arguments for treatment or compensation for treatment extend as well.

[93] Some might object that invoking the common good is unjustified because subjects of research are often motivated to participate for reasons other than an altruistic desire to serve the common good. This, however, would not vitiate the fact that whatever the motive of the subjects, they do, in fact, contribute to the common good through their participation in research and this fact is morally significant.

[94] Indeed, the DHEW went so far as to draft regulations requiring that institutions receiving funds from DHEW have in place "mechanisms to provide compensation for individuals who suffer injury as the result of their participation as subjects in biomedical or behavioral research," but these regulations were never enacted. DHEW. (1979). Draft Protection of Human Subjects: Proposed Regulations on Compensation of Human Subjects Injured in Biomedical and Behavioral Research.

[95] Neither body went beyond treatment or medical costs in their recommendations. NBAC. (2001, August). *Ethical and Policy Issues in International Research: Clinical Trials in Developing Countries*. Bethesda, MD; Mastroianni, A., Faden, R., and D. Federman. (Eds). Appendix D: Compensation Systems for Research Injuries. In Institute of Medicine, *Women and Health Research: Ethical and Legal Issues of Including Women in Clinical Studies, Vol. I*. Washington, DC: National Academy Press, pp. 243-252; Pike, E. (In press). Recovering from research: A No-fault proposal to compensate injured research participants. *American Journal of Law and Medicine*. Retrieved from http://papers.ssrn.com/sol3/papers.cfm?abstract_id=1923817; Presidential Commission for the Study of Ethical Problems in Medicine and Biomedical and Behavioral Research. (1982). *Compensating for Research Injuries*. Washington, DC: Department of Health, Education, and Welfare; Scott, L. (2003). Research-related injury: Problems and solution. *The Journal of Law, Medicine and Ethics*, (31)3, 421-422. (which touches on the history of compensation).

[96] Institute of Medicine (IOM). (2002). *Responsible Research: A Systems Approach to Protecting Research Participants*. Washington, DC: National Academies Press, p. 193. Act No. 885 of September 20, 2005 (as amended by Act No. 1545 of December 20, 2006 and Act No. 523 of June 6, 2007). The Liability for Damages Act. Copenhagen: Patientforsikringen. Retrieved from http://www.patientforsikringen.dk/en/Love-og-Regler/Lov-om-klage-og-erstatningsadgang/Behandlingsskader.aspx (accessed December 6, 2011); Act. No. 24 of January 21, 2009. The Right to Complain and Receive Compensation within the Health Service. Copenhagen: Patientforsikringen. Retrieved from http://www.patientforsikringen.dk/en/Love-og-Regler/Lov-om-klage-og-erstatningsadgang/Lægemiddelskader.aspx (accessed December 2, 2011); Council for International Organizations and Medical Sciences (CIOMS) and World Health Organization. (2002). *International Ethical Guidelines for Biomedical Research involving Human Subjects*. Geneva: World Health Organization; France. Code de la Santé Publique. Arts. L. 1121-7, L. 1121-10, L. 1142-22 (dedicated to the insurance and compensation of research subjects). Retrieved from http://www.legifrance.gouv.fr/ (accessed December 2, 2011); International Conference on Harmonisation. (1996, June). *Guideline for good clinical practice E6(R1)*. Retrieved from http://www.ich.org/fileadmin/Public_Web_Site/ICH_Products/Guidelines/Efficacy/E6_R1/Step4/E6_R1__Guideline.pdf; Ministry of Health of China. (2007). *Interim Measures for Guidelines on Ethical Review of Biomedical Research Involving Human Subjects*. Retrieved from http://www.chinaphs.org/bioethics/regulations_&_laws.htm#_Toc161968266 (accessed December 6, 2011); Uganda National Council for Science and Technology. (2007, March). *National Guidelines for Research Involving Humans as Research Participants*. Kampala, Uganda: UNCST; World Health Organization. (2005). *Handbook for good clinical research practice (GCP): guidance for implementation*; Many less developed nations also require treatment or compensation for treatment. These include: Brazil, India, the Philippines, and Uganda; Brazil. (1987). National Health Council. Resolution No. 196/96. Decree No. 93933 of January 14, 1987; Indian

Council of Medical Research. (2006). *Ethical Guidelines for Biomedical Research on Human Participants.* New Delhi, India: Director General of the Indian Council of Medical Research; Philippine Council for Health Research and Development. (2000). *National Guidelines for Biomedical/Behavioral Research.* Taguig, Metro Manila.

[97] PCSBI, *Research Across Borders,* op cit., p. 11. The Panel suggested that the "United States should implement a system to compensate research subjects for research-related injuries."

[98] Schaefer, G.O., Emanuel, E.J., and A. Wertheimer. (2009). The obligation to participate in biomedical research. *The Journal of the American Medical Association,* 302(1), 67-72.

[99] Nolan, J.R. and J.M. Nolan-Haley. (1990). *Deluxe Black's Law Dictionary: Sixth Edition.* St. Paul, MN: West Publishing Co.

[100] Weisbard, A.J. (1987). The role of philosophers in the public policy process: A view from the President's Commission. *Ethics,* 97(4), 780.

[101] Other examples where society does not permit informed volunteers to assume risks include the medical education context. Wikler, D., Professor of Ethics and Population Health, Harvard University. (2011). Compensation for Research-Related Injury. Presentation to PCSBI, November 17, 2011. Retrieved from http://bioethics.gov/cms/node/391 (accessed December 6, 2011).

[102] In other words, there are unforeseeable and/or unavoidable risks that are part of the nature of human subjects research and must be borne by someone if medical science is to progress. But it does not follow from this fact that good science and a social safety net are incompatible.

[103] The Commission does not endorse the view of research subjects as "wage-earners," however, because if volunteers were treated as "wage-earners," the research would violate the prohibitions on "undue inducement." 45 C.F.R. Part 46.

[104] Dickert, N. and C. Grady. (1999). What's the price of a research subject? Approaches to payment for research participation. *The New England Journal of Medicine,* 341(3), 198-203.

[105] This Commission goes further than past commissions in stating unequivocally that injured research subjects should not bear the financial costs of research injuries. After intensive study, the President's Commission for the Study of Ethical Problems in Medicine and Biomedical and Behavioral Research (1982) ultimately recommended only that "a small experiment be undertaken" to "determine the need for, and practical feasibility of, a compensation program"—a recommendation that was reiterated by ACHRE in 1995 and NBAC in 2001. No action was taken in response to this series of recommendations. ACHRE, op cit.; NBAC. (2001, August). *Ethical and Policy Issues in International Research: Clinical Trials in Developing Countries (2001).* Bethesda, MD: NBAC; Presidential Commission for the Study of Ethical Problems in Medicine and Biomedical and Behavioral Research. (1982). *Compensating for Research Injuries.* Washington, DC: Department of Health, Education, and Welfare.

[106] The University of Washington, for example, offers treatment to human research subjects injured overseas at their facility in Seattle or through reimbursement (restitution) to individuals who receive treatment overseas. Moe, K.E., Director and Assistant Vice Provost for Research, University of Washington. (2011). University of Washington Human Subjects Assistance Program. Presentation to PCSBI, November 17, 2011. Retrieved from http://bioethics.gov/cms/meeting-seven.

[107] In this regard, the Commission recognizes the research and analysis into questions of reparations and remedies produced by ACHRE. See ACHRE. (1995). *Final Report of the Advisory Committee on Human Radiation Experiments.* New York, NY: Oxford University Press, pp. 512-522 (providing an example of carefully parsed recommendations for different groups affected by the radiation experiments depending on their particular circumstances); Memorandum from Advisory Committee Staff to the Members of the Advisory Committee on Human Radiation Experiments. (December 12, 1994). Discussion Memo on Proposed Remedies and Guidelines. Retrieved from http://www.gwu.edu/~nsarchiv/radiation/dir/mstreet/commeet/meet9/brief9/tab_g/br9g1.txt (discussing possible remedies available for those affected by the human radiation experiments). See also Brooks, R.L., ed. (1999). *When Sorry Isn't Enough: The Controversy over Apologies and Reparations for Human Injustice.* New York: New York University (discussing a number of examples of reparations, and developing a "theory of redress" based on the concepts of human injustice and human rights); Randelzhofer, A., and C. Tomuschat, eds. (1999). *State Responsibility and the Individual: Reparation in Instances of Grave Violations of Human Rights.* Cambridge, MA: Kluwer Law International (addressing in particular claims for reparations under a several international covenants and agreements, including the European Convention on Human Rights); Schonsteiner, J. (2007). Dissuasive measures and the "society as a whole": A working theory of reparations in the Inter-American Court of Human Rights. *American University International Law Review,* 23, 127-164 (examining reparations as implemented by a particular international body).

[108] During the Tuskegee Syphilis Study, 600 African American men with syphilis were enrolled in a decades-long observational study of disease progression without receiving readily available and proven treatment. DHEW. (1973). *Final Report of the Tuskegee Syphilis Study Ad Hoc Advisory Panel.* Retrieved from http://biotech.law.lsu.edu/cphl/history/reports/tuskegee/tuskegee.htm (accessed December 6, 2011); Jones, J.H. (1993). *Bad Blood: The Tuskegee Syphilis Experiment.* New York, NY: The Free Press. Most of the actors in that infamous research are now deceased. Reverby, S. (2009). *Examining Tuskegee: The Infamous Syphilis Study and Its Legacy.* Chapel Hill, NC: University of North Carolina Press.

[109] Clinton, W.J. (1997, July 16). Remarks by the President in Apology for Study Done in Tuskegee. The White House. Washington, DC. Retrieved from http://clinton4.nara.gov/New/Remarks/Fri/19970516-898.html.

[110] PCSBI, *Research Across Borders,* op cit., p. 11.

[111] An individual injured as a result of research participation can sue for damages through the tort system. Claims may be brought on a range of intentional, negligence, strict liability and products-liability theories. See, e.g., Morreim, E.H. Medical research litigation and malpractice tort doctrines: Courts on a learning curve. *Houston Journal of Health Law and Policy,* 4, 1-86. Subjects may, for example, allege they did not give fully informed consent when exposed to certain adverse events from research, and thereby bring an intentional tort claim of battery against research sponsors. See, e.g., Morreim, E.H. (2004). Litigation in clinical research: Malpractice doctrines versus research realities, *Journal of Law, Medicine & Ethics,* 32, 474-481. Subjects could also allege negligence, which is the more typical malpractice-based action. To succeed on a civil claim of negligence, individuals who are injured must prove that the research sponsors violated a duty of reasonable care or a medical

standard of care—e.g., by failing to obtain fully informed consent or to minimize the risks of dangerous research—and that doing so caused and proximately caused physical or emotional injury. Ibid, p. 475; Pike, E. (In press). Recovering from Research: A No-Fault Proposal to Compensate Injured Research Participants. *American Journal of Law and Medicine*. Retrieved from http://papers.ssrn.com/sol3/papers.cfm?abstract_id=1923817; Injured research subjects may have difficulty recovering under strict liability; the restatement governing products liability for drugs and devices requires that risks of harm be foreseeable, a standard that injured research subjects may have difficulty satisfying. See Restatement (Third) of Torts: Prod. Liab. § 6(c) (1998) (strict products liability for drugs and devices turns on whether the risks of harm were foreseeable); Restatement (Third) of Torts: Prod. Liab. § 6, Comment g (1998) ("Imposing liability for unforeseeable risks can create inappropriate disincentives for the development of new drugs and therapeutic devices."). See also *Hufft v. Horowitz*, 4 Cal. App. 4th 8, 21 (1992) (arguing that the use of the strict liability standard with regards to pharmaceutical drugs and devices would be against the public interest because of the potential for stifling valuable biomedical research). Moreover, subjects injured as a result of participating in federally conducted research may have difficulty recovering due to other legal barriers. Pike, E. R. (In press). Recovering from Research: A No-Fault Proposal to Compensate Injured Research Participants. *American Journal of Law and Medicine*. Retrieved from http://papers.ssrn.com/sol3/papers.cfm?abstract_id=1923817

[112] VA. (2010). VHA Handbook 1200.05, para.31(a)(10)(b) and para. 60. October 15. The treatment guarantee is limited to research studies approved by a VA Research & Development Committee and conducted under the supervision of VA employees; it does not apply to injuries resulting from noncompliance by a subject or to research conducted for VA under a contract with an individual or non-VA institution.

[113] See 32 C.F.R. 108.4(i); DOD. (2002, March 25). Department of Defense Directive: Protections of Human Subjects and Adherence to Ethical Standards in DOD-supported Research 3216.02, sec. 5.3.4. DOD. (2007, December 3). Department of Defense Instruction 6000.08, sec. 6.2.4. DOD-Department of the Army. (1990, January 25). Army Regulation: Research and Development- Use of Volunteers as Subjects of Research 70-25, sec. 3-1(k); DOD-Department of the Army. (1989, September 1). Army Regulation: Medical Services--Clinical Investigation Program 40-38, sec. 3-3(j) and Appendix C; DOD-Department of the Navy. (2006, November 6). Human Research Protection Program SECNAV Instruction 3900.39D, sec. 6(a)(5); DOD-Office of the Under Secretary of Defense (Personnel and Readiness), Research Regulatory Oversight Office, Human Research Protection Program, Administrative Instruction, sec. 12.2.3, p.17; DOD-Air Force Research Laboratory. (2005, May 5). Air Force Instruction: Protection of Human Subjects in Research 40-402, sec. 3.3.2.

[114] OHRP. (2006). Sheet 6—Guidelines for Writing Informed Consent Documents. Retrieved from http://ohsr.od.nih.gov/info/sheet6.html (accessed December 6, 2011).

[115] The federal government also essentially self-insures for civil liability. The Federal Tort Claims Act was enacted in 1948. Under the FTCA, the U.S. Government can be found liable for a tort, "in the same manner and to the same extent as a private individual under like circumstances." The U.S. Government cannot, however, be liable for punitive damages. 28 U.S.C. § 2674.

[116] In 2010, Medicare provided guidance indicating that Medicare was consider the "secondary payer" in the event of other provision for compensation associated with clinical trial injury, and as a result, that compensation must be exhausted first before Medicare can be billed for care associated with injuries arising from clinical trials. Centers for Medicare and Medicaid Services. (n.d.). Medicare Coverage—Clinical Trials: Final National Coverage Decision. Retrieved from https://www.cms.gov/clinicalTrialPolicies/Downloads/finalnationalcoverage.pdf; Centers for Medicare and Medicaid Services. (2010, May 26). Clinical trials and liability insurance (including self-insurance), no-fault insurance, and workers' compensation. Retrieved from http://www.cms.gov/. Furthermore, Medicare covers routine costs arising from "qualifying clinical trials," such as conventional care that would be provided even if the patient were not enrolled in a clinical trial. Centers for Medicare and Medicaid Services (CMS). (n.d.). Medicare Coverage—Clinical Trials: Final National Coverage Decision. Retrieved from https://www.cms.gov/clinicalTrialPolicies/Downloads/finalnationalcoverage.pdf; CMS. (2010, May 26). Clinical Trials and Liability Insurance (Including Self-Insurance), No-Fault Insurance, and Workers' Compensation. Retrieved from http://www.cms.gov/MandatoryInsRep/Downloads/AlertClinicalTrailsNGHP.pdf. See also IOM. (2000). *Extending Medicare Reimbursement in Clinical Trials*. Washington, DC: National Academy Press.

[117] In the last five years, the program has written off approximately $250,000 in healthcare costs for care in the UW system and paid about $8,000/year in out-of-pocket compensation. Moe, K.E., Director and Assistant Vice Provost for Research, University of Washington. (2011). University of Washington Human Subjects Assistance Program. Presentation to the PCSBI, November 17. Retrieved from http://bioethics.gov/cms/meeting-seven (accessed December 6, 2011); The Committee on Human Research, The Human Research Protection Program, University of California San Francisco. (2006). *UCSF Guidance on Research Topics and Issues*. Retrieved from http://www.research.ucsf.edu/chr/Guide/chrH_Injury.asp (accessed December 6, 2011); Wake Forest University Health Sciences Institutional Review Board. (2007). *Research Related Injury*. Retrieved from http://www.wakehealth.edu/WorkArea/linkit.aspx?LinkIdentifier=id&ItemID=2126 (accessed December 6, 2011) (describing other institutional compensation programs).

[118] NASA provides compensation for subjects through its worker's compensation system in the event of injury related to intramural research (i.e., research performed at NASA, using NASA equipment or facilities, or where a NASA employee or on-site contractor serves as the principal investigator). See NASA. (2004). NASA Procedural Requirement 7100.1 – Protection of Human Research Subjects, secs.9.1.4, 11.6, and Appendix B, Revalidated July 7, 2008.

[119] The language in the consent form is as follows: "All forms of medical research, diagnosis, and treatment involve some risk of injury or illness. Despite our high level of precaution, you may develop an injury or illness due to participating in this study. If you develop an injury or illness determined by the on duty physician to be due to your participation in this research, the EPA will reimburse your medical expenses to treat the injury or illness up to $5000." See EPA. (2009, June). Guidance for Human Subjects Research in the National Exposure Research Laboratory, EPA 600/R-10/175, sec. 4.1.2.

ENDNOTES

[120] See NIH. (2011, October). *Grants Policy Statement*. Retrieved from http://grants.nih.gov/grants/policy/nihgps_2011/nihgps_2011.pdf (noting that "insurance usually is treated as an F&A [Facilities and Administration] cost. In certain situations, however, where special insurance is required as a condition of the grant because of risks peculiar to the project, the premium may be charged as a direct cost if doing so is consistent with organizational policy. Medical liability (malpractice) insurance is an allowable cost of research programs at educational institutions only if the research involves human subjects. If so, the insurance should be treated as a direct cost and assigned to individual grants based on the manner in which the insurer allocates the risk to the population covered by the insurance.").

[121] "During a subject's participation in a trial, the investigator . . . should ensure that reasonable medical care is provided to a subject for any adverse events, including clinically significant laboratory values, related to the trial participation. . . . Subjects should receive appropriate medical evaluation and treatment until resolution of any emergent condition related to the study intervention that develops during or after the course of their participation in a study, even if the follow-up period extends beyond the end of the study at the investigative site." FDA. (2009, October). Guidance for Industry, Investigator Responsibilities — Protecting the Rights, Safety, and Welfare of Study Subjects, p. 7. Retrieved from www.fda.gov/downloads/Drugs/GuidanceComplianceRegulatoryInformation/Guidances/UCM187772.pdf.

[122] Biotechnology Industry Organization (BIO). (2011). Comments submitted to PCSBI, October 28, 2011. ("In the private sector, sponsors are required to obtain clinical trial insurance to pay for medical expenses for injured trial participants."); Francer, J.K. Assistant General Counsel, PhRMA. (2011). Presentation to PCSBI, November 16. Retrieved from http://bioethics.gov/cms/node/389; Medford, R.M., Chairman and President of Salutria Pharmaceuticals. (2011). Presentation to PCSBI, November 16. Retrieved from http://bioethics.gov/cms/node/389.

[123] See Calabresi, G. (1961, March). Some thoughts on risk distribution and the law of torts. *Yale Law Journal,* 70, 449-553; Latin, H.A. (1985, May). Problem-solving behavior and theories of tort liability. *California Law Review,* 73, 677-746.

[124] The National Childhood Vaccine Injury Act of 1986, Public Law 99-660, 100 Stat. 3743. November 14, 1986; Federal Employees' Compensation Acts Amendments of 1966, Public Law 89-488, 80 Stat. 252. July 4, 1966.

[125] Air Transportation Safety and System Stabilization Act, Public Law 107-42, 115 Stat. 230. September 22, 2001 (establishing the 9/11 Fund).

[126] Neraas, M.B. (1988). The National Childhood Vaccine Injury Act of 1986: A solution to the vaccine liability crisis? *Washington Law Review,* 63, 149.

[127] For example, a varied system could be left open to a range of reasonable standards. The National Childhood Vaccine Injury Act sets forth two standards, one of which claimants must satisfy to obtain compensation: claimants requesting compensation for injuries listed in the Vaccine Injury Table have a rebuttable presumption of causation, while claimants requesting compensation for an off-Table injury must prove causation in fact by a preponderance of the evidence. See, e.g., Little, E.A. (2007). The role of special masters in off-table vaccination compensation cases: Assuring flexibility over certainty. *Federal Circuit Bar Journal,* 16, 356-57.

[128] Porter, J.P. and G. Koski, (2008). Regulations for the protection of humans in research in the United States: The Common Rule. In E.J. Emanuel, et al. (Eds). *The Oxford Textbook of Clinical Research Ethics*. New York, NY: Oxford University Press (pp. 156-167). (stating that "imposition of federal regulations to achieve these goals is not considered necessary or appropriate by some in the research community… the enthusiasm for these processes is dampened by the logistical, administrative, and regulatory burdens they necessarily entail, rather than by lack of concern for ethical conduct or well-being of research subjects.").

[129] Brock, W. (2008). Philosophical Justifications of Informed Consent in Research. In E.J. Emanuel, et al. (Eds). *The Oxford Textbook of Clinical Research Ethics*. New York, NY: Oxford University Press (pp. 606-612) (stating "there is always an inherent conflict of goals between researchers and participants, a conflict between participants' typical goal of securing their wellbeing and the researchers' goal of producing knowledge.").

[130] NIH promulgated new conflict of interest rules in the summer of 2011 citing a need based on the increased percentage of research funding by industry, complications created by multidisciplinary research teams that involve a larger number of researchers, and the increasingly complexity of growing ties between industry and academia. American Association of Medical Colleges and Association of American Universities. (2008). *Protecting Patients, Preserving Integrity, Advancing Health: Accelerating the Implementation of COI Policies in Human Subjects Research*; IOM. (2009). *Conflict of Interest in Medical Research, Education, and Practice*. Washington, DC: National Academies Press; Responsibility of Applicants for Promoting Objectivity in Research for which Public Health Service Funding is Sought and Responsible Prospective Contractors, 76 Fed. Reg. 53256 (Aug. 25, 2011). (to be codified at 42 C.F.R. pt. 50 and pt. 94).

[131] Beecher, H.K. (1966). Ethics and clinical research. *New England Journal of Medicine*, 274(24), 1354-1360.

[132] Prior to promulgation of the Common Rule and other human subjects protections, a number of incidents, some deeply disturbing and tragic, demonstrated that relying on investigators alone could not ensure a consistent baseline level of safety and respect for the interests of research subjects. In 1963, at the Jewish Chronic Disease Hospital in Brooklyn, 22 elderly patients were injected, without their consent, with live cancer cells. This protocol had no relation to the patients' medical care, and later investigations revealed that consent was not sought precisely because researchers wanted to avoid the possibility that patients would refuse. See Arras, J. (2011). The Jewish Chronic Disease Hospital Case: A Tragedy in Research Ethics. In E.J. Emanuel, et al. (Eds.). *The Oxford Textbook of Clinical Research ethics*. New York, NY: Oxford University Press (pp. 73-79); Coleman, C.H. et al (2005). *The Ethics and Regulation of Research with Human Subjects*. Newark, NJ: Lexis Nexis. p. 39; Between 1956 and 1971, a study at the Willowbrook State School sought to better understand the natural history of hepatitis, and the effects of gamma globulin in treating it, by injecting and inoculating the physically and cognitively impaired residents of the institution with the virus. Coleman, op cit. Later public scrutiny of this work raised deep concerns about the quality of consent provided, although, several prominent medical journals weighed in at the time commending the researchers for their "judicious use of human beings," and for benefitting the children by "infecting them under carefully controlled conditions and [giving] them expert attention." Coleman, op cit. (quoting The *Journal of the American Medical Association*

and the *New England Journal of Medicine*). Beecher was of course aware of the Brooklyn and Willowbrook cases and cited them in his landmark 1966 paper. Though he rejected "rigid rules" to govern investigator conduct, his own argument demonstrates the inadequacy of relying alone on the virtue of individual investigators. Though Beecher was wrong about the need for external constraints, he was decidedly right about the need for an internalized professional ethic. The ethical rationale the researchers voiced at the time was that these children would have become infected regardless because of the frequency of infection in the institution and lack of any known antidote. Similar and distressingly inadequate rationales were used to justify the infamous Tuskegee Study, in which impoverished African-American sharecroppers were misled and deliberately denied treatment for syphilis so that members of the U.S. Public Health Service could study the natural history of the disease. Reverby, S. (2009). *Examining Tuskegee: The Infamous Syphilis Study and Its Legacy*. Chapel Hill, NC: University of North Carolina Press.

[133] Through encouraging regulatory parsimony, the Commission recommended "only as much oversight as is truly necessary to ensure justice, fairness, security, and safety while pursuing the public good." PCSBI. (2010, December). *New Directions: The Ethics of Synthetic Biology and Emerging Technologies*. Washington, DC: PCSBI.

[134] PhRMA. (2011, July). PhRMA Principles on Conduct of Clinical Trials: Communication of Clinical Trial Results. Retrieved from http://www.phrma.org/sites/default/files/105/042009_clinical_trial_principles_final.pdf.

[135] For example, AstraZeneca has its own code of conduct. AstraZeneca. (2011, February). Code of Conduct. Retrieved from http://www.astrazeneca.com/Responsibility/Code-policies-standards.

[136] The Commission recognizes that the phrase "research profession" encompasses many different roles. Researchers may be trained in medicine, laboratory science, statistics or other fields, each of which has a unique professional culture that provides its members with varying instruction in research ethics. The disciplinary diversity endemic in the modern research enterprise underscores the need to develop shared professional norms for all who are trusted to do research on human subjects.

[137] As Atul Gawande discusses in *The New Yorker*: "There was a moment in sports when employing a coach was unimaginable—and then came a time when not doing so was unimaginable. We care about results in sports, and if we care half as much about results in schools and in hospitals we may reach the same conclusion…We could create coaching programs not only for surgeons but for other doctors, too—internists aiming to sharpen their diagnostic skills, cardiologists aiming to improve their heart-attack outcomes, and all of us who have to figure out ways to use our resources more efficiently. Gawande, A. (2011, October 3). Personal Best. *The New Yorker*. Some of the federal government funding agencies currently require ethics training for researchers. For example, NIH requires training in the protection of human subjects through NIH-based training or free web-based tutorial in human subjects research. NIH. (n.d.). Grants Policy Statement – Part II: Terms and Conditions of NIH Grant Awards, Subpart A, Chapter 4.1.15.5. Retrieved from http://grants.nih.gov/grants/policy/nihgps_2011/nihgps_2011.pdf.

[138] Geller, G., et al. (2010). Beyond "compliance": The role of institutional culture in promoting research integrity. *Academic Medicine*, 85(8), 1296-1302.

[139] For example, see Dr. Ronald Bayer's talk before the Commission. "I, like everyone else at the Mailman School of Public Health have to take an online test to guarantee that I have read the right things and understand the right things…I have to tell you, it is the most insulting experience to sit in front of a screen, to download a text and then a series of questions to which there is only one right answer, and if God forbid you think that there may be an ambiguity or an uncertainty, you get the answer wrong. They have to learn something, spit it back and give the right answer, and if you don't get a good enough score, you can't do research, you have to take the test again." Bayer, R., Professor of Sociomedical Sciences, Columbia University. (2011). Implementing Federal Standards – Ethics Issues. Presentation to PCSBI, May 19, 2011. Retrieved from http://bioethics.gov/cms/node/229 (accessed December 6, 2011).

[140] Arras, J.D. (1991, February). Getting down to cases: The revival of casuistry in bioethics. *Journal of Medicine and Philosophy*, 16(1), 29-51; Coutts, M.C. (1991). Teaching ethics in the health care setting. *Kennedy Institute of Ethics*, Scope Note 16. Retrieved from http://bioethics.georgetown.edu/publications/scopenotes/sn16.pdf; Macrina, F. L. and C. L. Munro. (1995). The case-study approach to teaching scientific integrity in nursing and the biomedical sciences. *Journal of Professional Nursing*, 11(1), 40-44. The Commission also plans to enlist experts to develop the Guatemalan case study into a teaching resource that could be used at the undergraduate level.

[141] PCSBI, *Research Across Borders*, op cit., p. 9.

[142] For example, between 1999 and 2003, Congress doubled the NIH's budget for research and development from a 1998 level of $13.6 billion to $27.3 billion. Office of Legislative Policy and Analysis. (n.d.). *Doubling the NIH Budget in the 107th Congress*. Retrieved from http://olpa.od.nih.gov/legislation/107/pendinglegislation/doubledec.asp (accessed December 6, 2011). In 2011, the NIH budget was $31.2 billion for research and development. While many of these funds are dedicated to research without human subjects, a proportionate increase has attached for human subjects research. See NIH. (n.d). *NIH Data Book*. Retrieved from http://report.nih.gov/nihdatabook (accessed December 6, 2011). See also, for privately funded research, Glickman, S. W., et al. (2009). Ethical and scientific implications of the globalization of clinical research. *The New England Journal of Medicine*, 360(8), 816-823; Office of the Inspector General. (2001). *The Globalization of Clinical Trials: A Growing Challenge in Protecting Human Subjects*. Boston, MA: HHS. Retrieved from http://oig.hhs.gov/oei/reports/oei-01-00-00190.pdf.

[143] For example, United Kingdom regulations delineate different requirements for continuing review than U.S. regulations do, but may protect research subjects just as well. See, e.g., Neaton, J., et al. (2010). Regulatory impediments jeopardizing the conduct of clinical trials in Europe funded by the National Institutes of Health. *Clinical Trials*, 7, 705-718.

[144] 45 C.F.R. § 46.101(a).

[145] 45 C.F.R. § 46.101(g).

[146] 45 C.F.R. § 46.101(h).

[147] PCSBI, *Research Across Borders*, op cit., p. 8.

ENDNOTES

[148] SACHRP to The Honorable Kathleen Sebelius. (October 13, 2011). Recommendations relative to the Department's recent Advance Notice of Proposed Rulemaking (ANPRM). Washington, DC; Letter from Barbara E. Bierer, Chair, SACHRP, to Kathleen Sebelius, Secretary, HHS (August 5, 2011). Retrieved from http://www.hhs.gov/ohrp/sachrp/commsec/commentspcsbi.pdf.pdf ("The lack of determinations of 'equivalence' – and of acceptable methods to determine 'equivalence' – has led to circumstances in which U.S.-based researchers and research institutions must insist on foreign entities' and foreign researchers' strict adherence to what can seem, to them, confusing and even impenetrable U.S. regulations and guidance documents. The solution is for the equivalent standard regulation to be implemented, as recommended by the Equivalent Protections Working Group, the National Bioethics Advisory Commission, and others.")

[149] PCSBI, *Research Across Borders*, op cit., p. 8.

[150] NBAC. (2001). *Ethical and Policy Issues in International Research: Clinical Trials in Developing Countries.* Bethesda, MD.: NBAC, p. 89.

[151] OIG. (2001). *The Globalization of Clinical Trials: A Growing Challenge in Protecting Human Subjects.* Boston, MA: HHS. p. 21. Retrieved from http://oig.hhs.gov/oei/reports/oei-01-00-00190.pdf.

[152] Equivalent Protections Working Group. (2003). *Report of the Equivalent Protections Working Group.* Washington, DC: HHS, p. 9. Retrieved from http://www.hhs.gov/ohrp/international/epwgreport2003.pdf.

[153] Protection of Human Subjects, Proposed Criteria for Determinations of Equivalent Protection. 70 Fed. Reg. 15,322-15,327 (March 25, 2005).

[154] Protection of Human Subjects: Interpretation of Assurance Requirements. 71 Fed. Reg. 38,645-38,646 (July 7, 2006).

[155] Davies, S., Director General of Research and Development and Chief Scientific Adviser for the Department of Health and NHS. (2011). Comments submitted to the PCSBI, May 13, 2011. Speaking with the Commission in March 2011, bioethicist Eric Meslin described the reaction in Canada to the failure to exercise equivalent protections authority. Meslin, E., Director, Indiana University Center for Bioethics. (2011). Social Justice and Ethics Issues. Presentation to the Presidential Commission of the Study of Bioethical Issues, March 1. Retrieved from http://bioethics.gov/cms/node/161. ("My own home country of Canada has an interesting conversation with the U.S. all the time about why it has to satisfy FDA and NIH rules when the Tri-Council policy is arguable as good as or better than [U.S. rules].")

[156] 21 C.F.R. § 312.120(a)(1); 21 C.F.R. § 814.15; FDA, *Guidance for Industry: Acceptance of Foreign Clinical Studies* (March 2001), Retrieved from http://www.fda.gov/downloads/RegulatoryInformation/Guidances/ucm124939.pdf (accessed December 2, 2011) (this guidance was superseded by 21 C.F.R. § 312.120 in 2008, but remains applicable to device clinical trials). FDA policy for foreign clinical trials is nuanced. FDA will not accept data from foreign studies that do not conform to good clinical practice in support of drug marketing applications, but it will examine such data. 21 C.F.R. §312.120(a)(2). Moreover, an applicant for marketing may seek a waiver of "any applicable requirements" (i.e., including the consent and independent review requirements) provided the applicant shows one of

the following: 1) why meeting the requirements is unnecessary or cannot be achieved, 2) an alternative course of action that satisfies the purpose of the requirement, or 3) other information that justifies a waiver. 21 C.F.R. § 312.120. The FDA, in turn, may grant a waiver "if it finds that doing so would be in the interest of the public health."

[157] It may also draw upon the experience of USAID, which recognizes that the standards established by the United Nations agencies, such as WHO, as "at least equivalent," and in guidance, sets forth three elements that, if met, could also result in a finding of equivalence. 22 C.F.R. §225.101(h); USAID. (2006, December 26). Protection of Human Subjects in Research Supported by USAID; A Mandatory Reference for ADS Chapter 200. Section 5. Retrieved from http://www.usaid.gov/policy/ads/200/200mbe.pdf (accessed December 2, 2011).

[158] Horowitz C.R., Robinson M., and S. Seifer. (2009). Community-based participatory research from the margin to the mainstream: Are researchers prepared? *Circulation*, 119(19), 2633-2842. (explaining that such research "provides a structure and mechanism for collaboration and rigorous research, using well-established or emerging methods, with a community focus" and that it "challenges researchers to list to, learn from, solicit and respect et contributions of, and share power, information and credit for accomplishments with the groups that they are trying to learn about and help.") at 2634; UNAIDS/AVAC. (2007). *Good Participatory Practice Guidelines for Biomedical HIV Prevention Trials*. UNAIDS: Geneva; PCSBI, *Research Across Borders*, op cit., p. 7; Clinical and Translational Science Awards Consortium and Community Engagement Key Function Committee Task Force on the Principles of Community Engagement. (2010). *Principles of Community Engagement 2d ed*. Bethesda, MD: NIH, pp. 7-17. Retrieved from http://www.atsdr.cdc.gov/communityengagement/.

[159] For further information on the Framingham Heart Study Ethics Advisory Board, see http://www.framinghamheartstudy.org/about/index.html (accessed December 6, 2011).

[160] For further information on the community advisory boards of the HIV Vaccine Trials Network, see http://www.hvtn.org/community/cab.html. See also National Institute of Allergy and Infectious Diseases (NIAID). (2009). *Community Recommendations Working Group of Community Partners and NIAID, Recommendations for Community Involvement in National Institute of Allergy and Infectious Diseases HIV/AIDS Clinical Trials Research*. Retrieved from http://www.hvtn.org/community/CAB_Recommendations_Certified.pdf (describing the requirements for community advisory boards in certain NIH-funded trials); NIH. (2004). *RFA AI-05-001: Leadership for NIH/AIDS Clinical Trials Networks*. Retrieved from http://grants.nih.gov/grants/guide/rfa-files/RFA-AI-05-001.html (program announcement detailing requirements for community advisory boards, community input and community representation to include training and education and resources required to meet community goals); Woodsong, C., and Q.A. Karim. (2005). A model designed to enhance informed consent: Experiences from the HIV Prevention Trials Network. *American Journal of Public Health*, 95(3), 412–419.

[161] Clinical and Translational Science Awards Consortium and Community Engagement Key Function Committee Task Force on the Principles of Community Engagement. (2010). *Principles of Community Engagement, 2d ed*. Bethesda, MD: NIH. Retrieved from http://www.atsdr.cdc.gov/communityengagement/.

[162] E.g., the RV144 trial discussed in Chapter 1.

[163] UNAIDS/WHO. (2007). *Ethical Considerations in Biomedical HIV Preventative Trials*. UNAIDS: Geneva; CIOMS. (2009). *International Ethical Guidelines for Epidemiological Studies*. CIOMS: Geneva; Nuffield Council on Bioethics. (2002, April). *The Ethics of Research Related to Healthcare in Developing Countries*, p. 77. Retrieved from http://www.nuffieldbioethics.org/sites/default/files/Ethics%20of%20research%20related%20to%20healthcare%20in%20developing%20countries%20I.pdf.

[164] PCSBI, *Research Across Borders*, op cit., p. 8.

[165] CIOMS. (2009). *International Ethical Guidelines for Epidemiological Studies*. Geneva: CIOMS; Nuffield Council on Bioethics. (2002, April). *The Ethics of Research Related to Healthcare in Developing Countries*. Retrieved from http://www.nuffieldbioethics.org/sites/default/files/Ethics%20of%20research%20related%20to%20healthcare%20in%20developing%20countries%20I.pdf.

[166] CIOMS. (2002). *International Ethical Guidelines for Biomedical Research Involving Human Subjects*. Geneva: CIOMS. p. 31.

[167] WMA. (2008). *The Declaration of Helsinki*. Ferney-Voltaire, France: WMA.

[168] UNAIDS; UNAIDS/AVAC. (2007). *Good Participatory Practice Guidelines for Biomedical HIV Prevention Trials*. Geneva: UNAIDS; UNAIDS/WHO. (2007). *Ethical Considerations in Biomedical HIV Preventative Trials*. Geneva.

[169] Australia National Health and Medical Research Council. (1991). *Guidelines on ethical matters in aboriginal and Torres Strait Islander research*. Canberra: Commonwealth of Australia. Retrieved from http://www.nhmrc.gov.au/guidelines/publications/e11; Medical Research Council of Canada, Natural Sciences and Engineering Council of Canada, and Social Sciences and Humanities Research Council of Canada. Chapter 9: Research Involving the First Nations, Inuit, and Métis peoples of Canada. In Social Sciences and Humanities Research Council of Canada. (2010). *Tri-Council Policy Statement: Ethical Conduct for Research Involving Humans*, 2d ed. Ottawa: Public Works and Government. Retrieved from http://www.pre.ethics.gc.ca/english/policystatement/introduction.cfm. Several of the Common Rule agencies also have policies related to community engagement. USAID requires both community involvement and community benefit for any USAID-supported research. USAID. (*Draft, unpublished*). Protection of Human Subjects In Research Supported by USAID. Attached to e-mail correspondence from Lee Claypool, Bureau for Global Health, USAID, to Valerie H. Bonham and Michelle Groman. (2011, June 23). The Indian Health Service (IHS) requires Area IRBs to, "determine and appraise the effect of research projects on the tribal organizations and Indian communities involved." Indian Health Service (IHS). (n.d.). *Indian Health Manual*. Part 1, ch.7, sec 1-7.5 (E) Retrieved from http://www.ihs.gov/ihm/index.cfm?module =dsp_ihm_pc_p1c7. In addition, the IRBs must assess the potential benefits of the proposal for the overall health of American Indians and Alaskan Natives.

[170] Lavery, J.V., et al. (2010). Towards a framework for community engagement in global health research" *Trends in Parasitology*, 26, 279-283; Tindana, P.O., et al. (2007). Grand challenges in global health: Community engagement in research in developing countries. *PLoS Medicine*, 4(9), e273; Weijer, C., and E. J. Emanuel. (2000). Protecting communities in biomedical research. *Science,* 289, 1142-1144.

171 Lavery, J.V., op cit.

172 UNAIDS/AVAC. (2011), op cit.

173 The National Commission for the Protection of Human Subjects of Biomedical and Behavioral Research. (1978). *The Belmont Report: Ethical Principles and Guidelines for the Protection of Human Subjects of Research*. Washington, DC: Department of Health, Education, and Welfare, DHEW Publication OS 78-0012, p. 9.

174 Ibid, p. 10.

175 See Wertheimer, A. (2008). Exploitation in clinical research. In Emanuel, E.J., et al. (Eds.). *The Oxford Textbook of Clinical Research Ethics*. New York, NY: Oxford University Press, pp. 201-210.

176 While there was no real effective treatment for syphilis when the study began in 1932, by 1943 it had been demonstrated that penicillin had the potential to be used as a curative agent for syphilis.

177 The Commission here focuses on the issues of justice. See Brandt, A.M. (1978). Racism and research: The case of the Tuskegee syphilis study. *The Hastings Center Report*, 8(6), 21-29; King, P.A. (1998). Race, justice, and research. In J.P. Kahn, A.C. Mastroianni, and J. Sugarman (Eds.). *Beyond Consent: Seeking Justice in Research*. New York: Oxford University Press, pp. 88-110. (both describing the injustice perpetrated in that case). But, it recognizes too that the deception and the lack of any informed consent represented grave ethical violations as well.

178 Robert M. Califf, M.D., Vice Chancellor for Clinical Research, Duke University, Professor of Medicine, Duke University Medical Center, Director, Duke Translational Medicine Institute. (2011). Presentation to the PCSBI, March 1. Retrieved from http://bioethics.gov/cms/meeting-four; Lawrence Corey, M.D., President and Director, Fred Hutchinson Cancer Research Center. Principal Investigator, HIV Vaccine Trials Network, Professor, Laboratory Medicine and Medicine, University of Washington. (2011). Presentation to the PCSBI, March 1. Retrieved from http://bioethics.gov/cms/meeting-four.

179 PCSBI. (2011, September). *Ethically Impossible*. Washington, DC: PCSBI, pp. 94-96. (discussing these fundamental principles).

180 Emanuel, E.J., et al. (2004). What makes clinical research in developing countries ethical? The benchmarks of ethical research. *The Journal of Infectious Diseases*, 189, 930-937; London, A.J. (2008). Responsiveness to host community health needs. In E.J. Emanuel, et al (Eds.). *The Oxford Textbook of Clinical Research Ethics*. New York, NY: Oxford University Press, pp. 737-744.

181 Council for International Organizations and Medical Sciences and World Health Organization. (2002). *International Ethical Guidelines for Biomedical Research involving Human Subjects,* Guideline 3 and Guideline 10. Geneva: World Health Organization. Guideline 3 states, in part, "The health authorities of the host country, as well as a national or local ethical review committee, should ensure that the proposed research is responsive to the health needs and priorities of the host country and meets the requisite ethical standards." Guideline 10 states, "Before undertaking research in a population or community with limited resources, the sponsor and the investigator must make every effort to ensure that: the research

is responsive to the health needs and the priorities of the population or community in which it is to be carried out; and any intervention or product developed, or knowledge generated, will be made reasonably available for the benefit of that population or community."

[182] WMA. (2008). *The Declaration of Helsinki*, Article 17. Ferney-Voltaire, France: WMA.

[183] NBAC. (2001). *Ethical and Policy Issues in International Research: Clinical Trials in Developing Countries*, Recommendation 1.3. Bethesda, MD: NBAC.

[184] Ibid; London, A.J., op cit.

[185] Ibid.

[186] Shah, S., Wolitz, R., and E.J. Emanuel. (2011). Refocusing the Responsiveness Requirement. *Bioethics*. doi: 10.1111/j.1467-8519.2011.01903.x.

[187] Just-in-time procedures allow specific elements of an application for an NIH grant to be submitted later in the application process, such elements may include certification of IRB approval if appropriate, or evidence of compliance with the education requirements to protect human research participants. See, e.g., NIH. (2011). *NIH Grants Policy Statement*, Section 2.5.1: Just-in-Time Procedures. Retrieved from http://grants.nih.gov/grants/policy/nihgps_2011/nihgps_ch2.htm#just_in_time_procedures; NSF. (2011). *Grant Proposal Guide, January 2011*. Retrieved from http://www.nsf.gov/pubs/policydocs/pappguide/nsf11001/gpgprint.pdf; Kelly, P.A., and M.L. Johnson. (2005). Just-in-time IRB review: Capitalizing on scientific merit review to improve human subjects research compliance. *IRB: Ethics and Human Research*, 27(2), 6-10.

[188] Glickman, S.W., et al. (2009). Ethical and scientific implications of the globalization of clinical research. *New England Journal of Medicine*, 360(8), 816-823.

[189] This section focuses on clinical research because it is the subset of research that most often raises concerns about study design. The principles and conclusions described herein may have utility for other fields, however, and the Commission reiterates its view that the fundamental ethical constraints on research with human subjects apply regardless of location, funding, or category. Implementation and understanding of ethical principles may vary across these spheres, however.

[190] For example, in 2000, Discovery Labs proposed a placebo-controlled Phase 3 trial to test the efficacy of a "me-too" surfactant (for the treatment of respiratory distress syndrome in premature infants) called Surfaxin in Latin America. Surfactants used for respiratory distress syndrome had been shown to decrease infant mortality by 34 percent; however none were available in the communities where the study was to be located. The infants in the active arm of the proposed trial were to receive Surfaxin, whereas all infants, including in the control arm, were to receive endotracheal tubes, ventilators, and antibiotics. In a case study on the topic, FDA official Robert Temple asked "why, if everyone in a trial is better off because of participation, and no one is denied anything otherwise available to them, the trial is not ethically acceptable?" Consumer advocates Peter Laurie and Sidney Wolfe countered that "[t]he ethical obligation of the researcher is to obtain needed scientific information in the manner most protective of the health of his or her study participants. Given the proven, life-saving effectiveness of other surfactants, a placebo-controlled trial could not satisfy this requirement, and an active-control trial was mandatory." Lavery, J.V., et al. (Eds). (2007).

Ethical Issues in International Biomedical Research: A Casebook. New York, NY: Oxford University Press, pp. 153-159.

[191] Ellenberg, S.S. and R. Temple. (2000). Placebo-controlled trials and active-control trials in the evaluation of new treatments: Part 1: Ethical and scientific issues. *Annals of Internal Medicine,* 133(6), 455-463. (arguing that "placebo-controlled trials may be ethically conducted even when effective therapy exists, as long as patients will not be harmed by participation and are fully informed about their alternatives"); Satcher, D. and H. Varmus. (1997). Ethical complexities of conducting research in developing countries. *New England Journal of Medicine* 337: 1003-1005.

[192] Angell, M. (1997). Ethics of clinical research in the Third World. *New England Journal of Medicine,* 337, 847-849; Lurie, P. and S. Wolfe. (1997). Unethical trials of interventions to reduce perinatal transmission of the human immunodeficiency virus in developing countries. *New England Journal of Medicine,* 337, 853-856.

[193] Appelbaum, P.S. (1996). Drug-free research in schizophrenia: An overview of the controversy. *IRB: A Review of Human Subjects Research,* 18(1), 1-5.

[194] While some have argued that challenge studies are per se unethical, see Weyers, W. (2003). *The Abuse of Man: An Illustrated History of Dubious Medical Experimentation*. New York, NY: Ardor Scribendi, p. 9. ("The best interest of individual human beings and the principle of preventing harm were never violated more openly than through experiments that employed inoculation, that is the deliberate infection of persons with pathogenic organisms for the expressed purpose of producing disease."), the Academy of Medical Sciences has argued that challenge studies are not unique in nature, ethically-speaking, from other study methods and may be conducted in an ethical manner under the same ethical standards that govern clinical drug trials. The Academy of Medical Sciences (2005, July). Microbial Challenge Studies of human volunteers. London: The Academy of Medical Sciences.

[195] The Commission recognizes that progress has been made by both scholars and the research community in carefully charting the contours of this middle ground, in a way that indicates not only where agreement has been forged but also where disagreements remain. Emanuel, E.J., and F.G. Miller. (2001). The ethics of placebo-controlled trials—a middle ground. *New England Journal of Medicine,* 345, 915-919; Kimmelman, J., Weijer, C., and E. Meslin. (2009). Helsinki discords: FDA, ethics, and international drug trials. *The Lancet,* 373(9657), 13-14. Speaking with the Commission in March 2011, philosopher Dan Brock supported a middle ground driven by professional mores in this arena. ("[T]he example that we have talked about, the standard of care example, is one where I am not sure who has the authority to make the law but there could be developed a consensus on a process to evaluate the standard of care in different countries and then a consensus that if Merck wants to come in or Pfizer wants to come in and do studies there, it has to meet this middle ground that we were talking about.") Brock, D., Director, Division of Medical Ethics, Harvard Medical School. (2011). Social Justice and Ethics Issues. Presentation to the Presidential Commission of the Study of Bioethical Issues, March 1. Retrieved from http://bioethics.gov/cms/node/161.

[196] Emanuel, E.J. and F.G. Miller, op cit.

[197] CIOMS. (2002). *International Ethical Guidelines for Biomedical Research Involving Human Subjects,* Commentary on Guideline 11, pp. 56-59. Retrieved from http://www.cioms.ch/

publications/layout_guide2002.pdf; NBAC. (2001, August). *Ethical and Policy Issues in Research Involving Human Participants*, Recommendation 2.2 and 28. Bethesda, MD: NBAC.

[198] WMA. (2008). *The Declaration of Helsinki*. Ferney-Voltaire, France: WMA. The 2008 Declaration permits some exceptions. It requires that "the benefits, risks, burdens and effectiveness of a new intervention must be tested against those of the best current proven intervention, except in the following circumstances: (1)The use of placebo, or no treatment, is acceptable in studies where no current proven intervention exists; or (2) Where for compelling and scientifically sound methodological reasons the use of placebo is necessary to determine the efficacy or safety of an intervention and the patients who receive placebo or no treatment will not be subject to any risk of serious or irreversible harm. Extreme care must be taken to avoid abuse of this option." See Ibid, Section C(32). By way of background, in 2000, the Declaration stated that a new method of treatment should be tested against the "best current…methods" of treatment. (There is also a consistent exception in the Declaration for when no method of intervention exists.) Concerns were raised that requiring the "best current" method of treatment in all studies was over-inclusive and would rule out many important clinical trials. For these reasons, in 2002, the WMA approved a "Note of Clarification" to their position on placebo-controlled trials stating that a placebo-controlled trial might be ethically acceptable if placebo was necessary for "compelling and scientifically sound methodological reasons," or the method under investigation was for a minor condition and those subjects receiving placebo would not be exposed to "any additional risk of serious or irreversible harm."

[199] Gutmann, A., PCSBI Chair. PCSBI Meeting 7, November 16, 2011. Retrieved from http://tvworldwide.com/events/bioethics/111116; Hyder, A.A., et al. (2004). Ethical review of health research: A perspective from developing country researchers. *Journal of Medical Ethics*, 30, 68-72; Lemmens, T. (1999). In the name of national security: Lessons from the Final Report on the Human Radiation Experiments. *European Journal of Health Law*, 6(7), 7-23.

[200] Angell, M., op cit.

[201] See Walach, H. et al. (2006) for the difference between "equipoise" and "clinical equipoise:" "[e]quipoise…means that there is no preference based on systematic knowledge for a treatment over an alternative or no treatment. Clinical equipoise…refers to the notion that there is honest disagreement about the optimal treatment among the medical community or between important sectors of the community. Walach, H., et al. (2006). Circular instead of hierarchical: methodological principles for the evaluation of complex interventions. *BMC Medical Research Methodology*, 6, 29.

[202] Commentators have differing views not only about the ethical source of clinical equipoise but also about its scope or the nature of its demands.

[203] Freedman, B. (1990). Placebo-controlled trials and the logic of clinical purpose. *IRB: A Review of Human Subjects Research*, 12(6), 5-6.

[204] London, A. J. (2007). Clinical equipoise: Foundational requirement or fundamental error? In Steinbock, B. (Ed.) *The Oxford Handbook of Bioethics*. New York, NY: Oxford University Press, pp. 571-596; Kukla, R. (2007). Resituating the principle of equipoise: Justice and access to care in non-ideal conditions. *Kennedy Institute of Ethics Journal*, 17(3), 171-202.

[205] In their view, an active study design comparing the already validated longer course of AZT treatment (the 076 protocol) against the shorter (and less expensive) course designed for deployment in the developing world was the only ethical option. See Angell, M. op cit.; Lurie, P. and S. Wolfe, op cit.; Connor, E.M, et al. (1994). Reduction of maternal-infant transmission of human immunodeficiency virus type 1 with zidovudine treatment. *New England Journal of Medicine*, 331, 1173-80. By contrast, Alex London believes that the critics of the short course AZT trial harbor an excessively narrow conception of what factors should be held in equipoise (i.e. just mortality data or a larger 'portmanteau' conception of equipoise that would include local differences such as sustainability, culture, or genetics). London, A. J. (2007). Clinical equipoise: Foundational requirement or fundamental error? In Steinbock, B. (ed) *The Oxford Handbook of Bioethics*. New York, NY: Oxford University Press, pp. 571-596.

[206] Hearkening back to Aristotle's famous dictum that those who are equal in relevant respects should receive equal treatment, these critics argued that giving human subjects from poor countries a placebo while their better off counterparts in Europe and the United States received the full course of AZT constituted a profound moral injustice. In the critics' view, the only theoretical perspective that could possibly justify such disparate treatment of patients suffering from the same condition was a dubious "moral relativism" according to which what's true ethically for people at one time and place might differ radically from what's ethically true for people in other circumstances. According to such a view, people in rich countries might maintain very high standards of appropriate treatment for research subjects, standards which were "true for them," while people in poor countries might well embrace much lower standards of acceptable treatment that were true for them. Another way to put this objection is to say that mere differences in socio-economic status should not count as morally relevant considerations in the design of clinical trials. See Macklin, R., Professor of Bioethics, Department of Epidemiology and Population Health, Albert Einstein College of Medicine. (2011). Presentation to PCSBI, August 30, 2011. Retrieved from http://bioethics.gov/cms/node/321; Angell, M. (1988). Ethical imperialism?: Ethics in international collaborative clinical research. *New England Journal of Medicine,* 319, 1081-1083; Annas, G. J. and M. A. Grodin. (1998). Human rights and maternal-fetal HIV transmission prevention in Africa. *American Journal of Public Health*, 88(4), 560-563.

[207] See *Justifying Site Selection* section above.

[208] See *Justifying Site Selection* section above.

[209] Many of these population-specific and infrastructure deficiencies were on display in the midst of the original perinatal AZT trials. The global state of the art, the 076 protocol, called for intravenous delivery of the drug not just at the time of birth but also well before and well after. This was an enormous problem in poor African countries where the first contact between pregnant women and the health system would usually take place just shortly before birth. That protocol also assumed that the women under study would not be breastfeeding their newborns and would instead be relying on infant formula, but this proviso would have forced these mothers to dilute the formula powder with water from contaminated local sources, thus placing their newborn children at grave risk for life-threatening diarrhea. And there were also legitimate worries expressed at the time that AZT might well have harmful side effects on women who were already plagued with iron deficiencies. On top of all these concerns, there was the fact that the 076 protocol, then currently costing roughly $1,000 per case, was obviously unsustainable economically in the countries hosting the trial, most

of which spent less than $10 per person per year from their meager public health budgets. See Dabis, F., et al. (1999). 6-month efficacy, tolerance, and acceptability of a short regimen of oral zidovudine to reduce vertical transmission of HIV in breastfed children in Côte d'Ivoire and Burkina Faso: a double-blind placebo-controlled multicentre trial. DITRAME Study Group. Diminution de la Transmission Mère-Enfant. *Lancet*, 353, 786-792; Crouch, R.A., and J.D. Arras. (1998). AZT trials and tribulations. *Hastings Center Report*, 28, 26-34; Wendler, D., et al. (2004). The standard of care debate: Can research in developing countries be both ethical and responsive to those countries' health needs? *American Journal of Public Health*, 94, 923-928.

[210] London, A. J. (2000). The ambiguity and the exigency: Clarifying 'standard of care' arguments in international research. *Journal of Medicine and Philosophy*, 25(4), 379-385.

[211] The inadequacy of this purported standard is amply demonstrated by the fact that the U.S. Public Health Service researchers could have cited it as a justification for giving no treatment to poor African-American subjects in the infamous Tuskegee Study, for whom medical treatment for syphilis was made unavailable, both before and after the advent of penicillin.

[212] Bhutta, Z. A. (2002). Ethics in international health research: A perspective from the developing world. *Bulletin of the World Health Organization*, 80, 117-120.

[213] See Hellman, S., and D.S. Hellman. (1991). Of mice and not men: Problems of the randomized clinical trial. *New England Journal of Medicine*, 324, 1585-1589; Relton, C., et al. (2010). Rethinking pragmatic randomised controlled trials: Introducing the "cohort multiple randomised controlled trial" design. *British Medical Journal*, 340, c1066; Truog, R.D. (1992). Randomized clinical trials: Lessons from ECMO. *Clinical Research*, 40(3), 519-522.

[214] See "A well-designed study that shows superiority of a treatment to a control (placebo or active therapy) provides strong evidence of the effectiveness of the new treatment, limited only by the statistical uncertainty of the result. No information external to the trial is needed to support the conclusion of effectiveness." Ellenberg, S.S., and R. Temple. (2000). Placebo-controlled trials and active-control trials in the evaluation of new treatments: Part 1: Ethical and scientific issues. *Annals of Internal Medicine*, 133(6), 455-463.

[215] For example, researchers run a trial designed to test the equivalence of drug A (the established standard treatment worldwide) against drug B (a new drug that is hopefully as effective but also less expensive, or with fewer side effects, etc) In this trial, drug B proves to be as effective as drug A in a head to head comparison without a placebo control. Although the intuitive conclusion to draw is that both A and B are equally effective, this might reflect a false positive error. Unless there are good reasons to believe that drug A, the established drug, was indeed effective in the specific population under study in the trial, it could be that neither drug was effective—i.e., that both were equally ineffective. Temple, R., and S.S. Ellenberg, op cit.

[216] Makuch, R., and M. Johnson. (1989). Issues in planning and interpreting active control equivalence studies. *Journal of Clinical Epidemiology*, 42, 503-511; Senn, S. (1993). Inherent difficulties with active control equivalence studies. *Statistics in Medicine* 12(24), 2367-2375; Temple, R. and S.S. Ellenberg, op cit.

[217] Anderson, J.R. (2001). Perinatal transmission and HIV: An unfinished success story. *The Hopkins HIV Report*, pp. 2-6.

[218] Emanuel, E. and F.G. Miller. (2001). The ethics of placebo controlled trials: A middle ground. *New England Journal of Medicine*, 345, 915-919; Cubeddu, L.X., et.al. (1990). Efficacy of ondansetron (GR 38032F) and the role of serotonin in cisplatin-induced nausea and vomiting. *New England Journal of Medicine*, 322, 810-860.

[219] Human Subjects Research Protections: Enhancing Protections for Research Subjects and Reducing Burden, Delay, and Ambiguity for Investigators, 76 Fed. Reg. 143, 44,512 (July 26, 2011).

[220] In addition, the Commission submitted comments to OSTP and the Secretary on October 24, 2011. Amy Gutmann, Chair, and James Wagner, Vice-Chair, to the Honorable John Holdren, Assistant to the President for Science and Technology and Director, OSTP, and the Honorable Kathleen Sebelius, Secretary, HHS. October 24, 2011.

[221] 45 C.F.R. § 46.101.

[222] 45 C.F.R. § 46.110.

[223] 45 C.F.R. § 46.101.

[224] OHRP. (2003, August). *Guidance on Expedited Review Procedures*. Retrieved from http://www.hhs.gov/ohrp/policy/exprev.html.

[225] Human Subjects Research Protections: Enhancing Protections for Research Subjects and Reducing Burden, Delay, and Ambiguity for Investigators, 76 Fed. Reg. 143, 44,513 (July 26, 2011).

[226] Ibid.

[227] 45 C.F.R. § 46.110.

[228] Brako, L., Assistant Vice President, University of Michigan, et al. (2011). Federal Demonstration Partnership (FDP) Current Initiatives for Reducing Burden of Regulatory Compliance [Powerpoint]. Presentation to the Secretary's Advisory Committee on Human Research Protections, March 8, 2011. Retrieved from http://www.hhs.gov/ohrp/sachrp/mtgings/mtg03-11/20110308fdp_.pdf. The FDP is a cooperative initiative consisting of 10 federal agencies and approximately 120 other recipients of federal funds (mostly academic institutions).

[229] 45 C.F.R. § 46.114.

[230] Human Subjects Research Protections: Enhancing Protections for Research Subjects and Reducing Burden, Delay, and Ambiguity for Investigators, 76 Fed. Reg. 143, 44,513 (July 26, 2011).

[231] 45 C.F.R. § 46.116.

[232] Human Subjects Research Protections: Enhancing Protections for Research Subjects and Reducing Burden, Delay, and Ambiguity for Investigators, 76 Fed. Reg. 143, 44,513 (July 26, 2011).

[233] A recent example of discord between the adaptation of the Common Rule among differing agency signatories is EPA's recent proposed rule on human subjects research involving pesticides, a final rule for which is due out in December 2011. Revisions to EPA's Rule on Protections for Subjects in Human Research Involving Pesticides. 76 Fed. Reg. 22, 5735,

ENDNOTES IV

5755 (Feb. 2, 2011). See also, Sean E. Peters, U.S. Nuclear Regulatory Commission, to Valerie Bonham (2011, November 9). E-mail Correspondence.

[234] For example, the Nuclear Regulatory Commission funds human subjects research but has not signed onto the Common Rule. Letter from Brian W. Sheron, Director, Office of Nuclear Regulatory Research, NRC, to Valerie H. Bonham, Executive Director, PCSBI. (November 9, 2011). Scientific Studies Supported by the U.S. Nuclear Regulatory Commission.

[235] Human Subjects Research Protections: Enhancing Protections for Research Subjects and Reducing Burden, Delay, and Ambiguity for Investigators, 76 Fed. Reg. 143, 44,513 (July 26, 2011).

[236] PCSBI. (2011, August). *Research Across Borders*. Washington, DC: PCSBI, p. 11.

[237] Human Subjects Research Protections: Enhancing Protections for Research Subjects and Reducing Burden, Delay, and Ambiguity for Investigators, 76 Fed. Reg. 143, 44,513 (July 26, 2011).

[238] Ibid.

[239] PCSBI, *Research Across Borders*, op cit., p. 10.

[240] In 1977, the Task Force on the Compensation of Injured Research Subjects produced a report finding that subjects involved in research supported by the U.S. Public Health Service should be compensated for injury going above and beyond that caused by any underlying illness that the subject might be experiencing. Secretary's Task Force on the Compensation of Injured Research Subjects. (1977). *Final Report*. Washington, D.C.: Department of Health, Education, and Welfare. HEW Secretary Caspar Weinberger convened the Task Force in 1975 in response to a memorandum from Theodore Cooper, M.D., Acting Assistant Secretary for Health, which outlined several ways in which the issue of compensation for research-related injury might be addressed. Memorandum from Theodore Cooper, Acting Assistant Secretary for Health, to Caspar Weinberger, Secretary of DHEW. February 25, 1975 (describing possible responses including: requiring grantees and contractors to provide assurance that they would compensate subjects injured in the course of research; establishing a no-fault compensation fund to compensate subjects for research-related injury; defining research subjects as federal employees for the purposes of the Federal Employees Worker's Compensation Act, 5 U.S.C. 8101 et seq.; and developing a federal "re-insurance" program that would provide assistance to institutions that were insured or self-insured if their liability went over a certain fixed amount). See also the Presidential Commission for the Study of Ethical Problems in Medicine and Biomedical and Behavioral Research. (1982). *Compensating for Research Injuries*. Washington, D.C.: U.S. Government Printing Office (recommending that HHS conduct a pilot study to determine the need for and feasibility of creating a program to provide compensation for injured research subjects).

APPENDICES

MORAL SCIENCE Protecting Participants in Human Subjects Research

Appendix I: Human Subjects Research Landscape Project: Scope and Volume of Federally Supported Human Subjects Research

Table Title

I.1	Departments/Agencies and HHS Operating Divisions
I.2	Extramural Human Subjects Projects Over Time
I.3	Intramural Human Subjects Projects Over Time
I.4	HHS Human Subjects Projects Over Time
I.5	HHS Extramural Human Subjects Projects Over Time
I.6	HHS Intramural Human Subjects Projects Over Time
I.7	Location of Human Subjects Projects Over Time
I.8	Location of HHS Human Subjects Projects Over Time
I.9	Total Extramural Award Funding (Human Subjects Projects) Over Time
I.10	Total HHS Extramural Award Funding (Human Subjects Projects) Over Time
I.11	Top 20 (of 117) Human Subjects Project Site Countries by Number of Projects, FY10
I.12	Top 20 (of 68) Extramural Awardee Institution Countries by Number of Human Subjects Projects, FY10
I.13	Location of Unique Extramural (Human Subjects Projects) Awardee Institutions, FY10
I.14	Proportion of Extramural Human Subjects Projects with Known Award Amounts
I.15	Proportion of Human Subjects Projects with Known Site Country
I.16	Proportion of HHS Human Subjects Projects with Known Site Country
I.17	Proportion of Human Subjects Projects with Known Subject Count
I.18	Proportion of Human Subjects Projects with Known Exempt/Non-exempt Status
I.19	Proportion of Human Subjects Projects with Known Number of Sites

APPENDIX I: Scope and Volume of Federally Supported Human Subjects Research

Table I.1 Department/Agencies and HHS Operating Divisions

Departments and Agencies that Received Data Request

DEPARTMENT/AGENCY	ABBREVIATION
Agency for International Development	USAID
Central Intelligence Agency	CIA
Consumer Product Safety Commission	CPSC
Department of Agriculture	USDA
Department of Commerce	DOC
Department of Defense	DOD
Department of Education	ED
Department of Energy	DOE
Department of Health and Human Services	HHS
Department of Homeland Security	DHS
Department of Housing and Urban Development	HUD
Department of Justice	DOJ
Department of Transportation	DOT
Department of Veterans Affairs	VA
Environmental Protection Agency	EPA
National Aeronautics and Space Administration	NASA
National Science Foundation	NSF
Social Security Administration	SSA

HHS Operating Divisions

UNIT	ABBREVIATION
Agency for Healthcare Research and Quality	AHRQ
Assistant Secretary for Preparedness and Response	ASPR
Centers for Disease Control and Prevention	CDC
Centers for Medicare & Medicaid Services	CMS
Food and Drug Administration	FDA
Health Resources and Services Administration	HRSA
Indian Health Service	IHS
National Institutes of Health	NIH
National Vaccine Program Office	NVPO
Office of Adolescent Health	OAH
Office of Population Affairs	OPA
Substance Abuse and Mental Health Services Administration	SAMHSA

Table I.2 Extramural Human Subjects Projects Over Time[†]

DEPARTMENT/ AGENCY	FY06	FY07	FY08	FY09	FY10	MEAN
HHS	21,676	21,491	20,929	22,309	22,322	21,745
DOD	2,976	2,946	2,925	1,890	2,107	2,569
NSF	1,820	2,271	2,627	2,988	3,051	2,551
ED	*222*	*1,145*	*1,199*	*1,296*	*1,969*	*1,166*
USDA	68	75	178	167	231	144
DOJ	95	107	74	125	188	118
USAID	159	157	106	64	62	110
DOE	88	70	72	60	52	68
EPA	29	29	35	17	40	30
DHS[‡]	N/R	19	30	23	10	22
NASA	8	15	17	23	26	18
SSA	10	18	17	14	13	14
HUD	7	11	7	18	18	12
DOT	*4*	*8*	*11*	*18*	19	*12*
DOC	11	8	13	10	10	10
CPSC	0	0	0	0	1	0
VA	0	0	0	0	0	0
CIA[§]	N/R	N/R	N/R	N/R	N/R	
Total	27,173	28,370	28,240	29,022	30,119	

[†] "Extramural" includes projects conducted solely extramurally and projects with an extramural component. "Projects" include awards and individual studies. "N/R" means that the data were not reported to the Commission. Departments/agencies that appear italicized reported that they were unable to provide complete data. See Appendix II for additional detail.
[‡] DHS reported that there "are no earlier data" than FY07. Mean is of reported years.
[§] The CIA did not submit project-level data to the Commission's database because these data are confidential (although not classified).

APPENDIX I: Scope and Volume of Federally Supported Human Subjects Research

Table I.3 Intramural Human Subjects Projects Over Time[†]

DEPARTMENT/ AGENCY	FY06	FY07	FY08	FY09	FY10	MEAN
VA	16,763	16,731	16,706	16,383	15,415	16,400
HHS	3,599	4,209	4,239	4,203	4,329	4,116
DOD	3,542	3,611	3,961	4,389	4,977	4,096
DOE	319	298	299	288	311	303
NASA	75	89	104	110	110	98
USDA	50	47	42	34	41	43
DOT	*22*	*18*	*24*	*29*	*37*	*26*
DOJ	25	11	12	15	28	18
DOC	10	18	21	17	13	16
EPA	7	12	8	4	6	7
CPSC	1	1	2	1	0	1
DHS[‡]	N/R	3	0	1	0	1
SSA	5	0	0	0	0	1
USAID	0	0	0	0	0	0
ED	*0*	*0*	*0*	*0*	*0*	*0*
HUD	0	0	0	0	0	0
NSF	0	0	0	0	0	0
CIA[§]	N/R	N/R	N/R	N/R	N/R	N/R
Total	24,418	25,048	25,418	25,474	25,267	

[†] "Projects" include awards and individual studies. "N/R" means that the data were not reported to the Commission. Departments/agencies that appear italicized reported that they were unable to provide complete data. See Appendix II for additional detail.
[‡] DHS reported that there "are no earlier data" than FY07. Mean is of reported years.
[§] The CIA did not submit project-level data to the Commission's database because these data are confidential (although not classified).

Table I.4 HHS Human Subjects Projects Over Time†

UNIT	FY06	FY07	FY08	FY09	FY10	MEAN
NIH	22,396	23,210	22,709	24,028	23,891	23,247
CDC	1,739	1,387	1,380	1,334	1,317	1,431
AHRQ	668	634	598	655	898	691
IHS	254	277	238	239	251	252
FDA	90	103	122	192	200	141
HRSA	47	51	50	43	35	45
OAH‡	x	x	x	x	31	31
NVPO	61	14	39	0	0	23
OPA	12	16	22	17	16	17
ASPR	8	8	10	4	12	8
CMS	0	0	0	0	0	0
SAMHSA	0	0	0	0	0	0
Total	25,275	25,700	25,168	26,512	26,651	

† "Projects" include awards and individual studies. See Appendix II for additional detail.
‡ OAH was established in 2010. Mean is of reported years.

APPENDIX I: Scope and Volume of Federally Supported Human Subjects Research V

Table I.5 HHS Extramural Human Subjects Projects Over Time[†]

UNIT	FY06	FY07	FY08	FY09	FY10	MEAN
NIH	20,632	20,526	19,922	21,255	21,006	20,668
AHRQ	562	550	517	574	830	607
CDC	277	251	289	310	304	286
FDA	67	66	71	93	74	74
HRSA	44	47	46	40	31	42
OAH[‡]	x	x	x	x	31	31
NVPO	61	14	39	0	0	23
OPA	12	16	22	17	16	17
IHS	13	13	13	16	18	15
ASPR	8	8	10	4	12	8
CMS	0	0	0	0	0	0
SAMHSA	0	0	0	0	0	0
Total	21,676	21,491	20,929	22,309	22,322	

[†] "Extramural" includes projects conducted solely extramurally and projects with an extramural component. "Projects" include awards and individual studies. See Appendix II for additional detail.
[‡] OAH was established in 2010. Mean is of reported years.

Table I.6 HHS Intramural Human Subjects Projects Over Time†

UNIT	FY06	FY07	FY08	FY09	FY10	MEAN
NIH	1,764	2,684	2,787	2,773	2,885	2,579
CDC	1,462	1,136	1,091	1,024	1,013	1,145
IHS	241	264	225	223	233	237
AHRQ	106	84	81	81	68	84
FDA	23	37	51	99	126	67
HRSA	3	4	4	3	4	4
ASPR	0	0	0	0	0	0
CMS	0	0	0	0	0	0
NVPO	0	0	0	0	0	0
OAH‡	x	x	x	x	0	0
OPA	0	0	0	0	0	0
SAMHSA	0	0	0	0	0	0
Total	3,599	4,209	4,239	4,203	4,329	

† "Projects" include awards and individual studies. See Appendix II for additional detail.
‡ OAH was established in 2010. Mean is of reported years.

APPENDIX I: Scope and Volume of Federally Supported Human Subjects Research   V

Table I.7 Location of Human Subjects Projects Over Time†

DEPARTMENT/ AGENCY	LOCATION	FY06	FY07	FY08	FY09	FY10
CIA‡	Domestic	N/R	N/R	N/R	N/R	N/R
	Foreign	0	0	0	0	0
	Mixed	0	0	0	0	0
	Unknown	0	0	0	0	0
CPSC	Domestic	1	1	2	1	1
	Foreign	0	0	0	0	0
	Mixed	0	0	0	0	0
	Unknown	0	0	0	0	0
DHS§	Domestic	N/R	20	26	23	10
	Foreign	N/R	2	4	1	0
	Mixed	N/R	0	0	0	0
	Unknown	N/R	0	0	0	0
DOC	Domestic	21	26	34	27	23
	Foreign	0	0	0	0	0
	Mixed	0	0	0	0	0
	Unknown	0	0	0	0	0
DOD	Domestic	2,451	4,904	4,775	5,082	6,787
	Foreign	203	379	368	255	265
	Mixed	2	6	5	8	30
	Unknown	3,862	1,268	1,738	934	2
DOE	Domestic	391	356	356	334	348
	Foreign	14	10	13	12	14
	Mixed	2	2	2	2	1
	Unknown	0	0	0	0	0

† "Projects" include awards and individual studies. "Mixed" means projects with both domestic and foreign components. "N/R" means that the data were not reported to the Commission. Departments/agencies that appear italicized reported that they were unable to provide complete data. See Appendix II for additional detail.
‡ The CIA did not submit project-level data to the Commission's database because these data are confidential (although not classified), but the agency did advise the Commission that all of its human subjects research takes place in the United States.
§ DHS reported that there "are no earlier data" than FY07.

continued

Table I.7 Location of Human Subjects Projects Over Time[†]

DEPARTMENT/ AGENCY	LOCATION	FY06	FY07	FY08	FY09	FY10
DOJ	Domestic	118	118	86	140	215
	Foreign	0	0	0	0	1
	Mixed	0	0	0	0	0
	Unknown	2	0	0	0	0
DOT	Domestic	26	26	35	47	56
	Foreign	0	0	0	0	0
	Mixed	0	0	0	0	0
	Unknown	0	0	0	0	0
ED	Domestic	114	1,067	1,104	1,151	1,538
	Foreign	0	6	30	65	115
	Mixed	0	0	0	0	0
	Unknown	108	72	65	80	316
EPA	Domestic	32	39	37	20	44
	Foreign	4	2	6	1	2
	Mixed	0	0	0	0	0
	Unknown	0	0	0	0	0
HHS	Domestic	24,176	23,173	22,303	23,633	23,539
	Foreign	370	401	474	503	481
	Mixed	654	854	1085	1,375	1,577
	Unknown	75	1,272	1,306	1,001	1,054
HUD	Domestic	7	11	7	18	18
	Foreign	0	0	0	0	0
	Mixed	0	0	0	0	0
	Unknown	0	0	0	0	0
NASA	Domestic	74	104	120	132	133
	Foreign	0	0	1	1	1
	Mixed	0	0	0	0	0
	Unknown	9	0	0	0	2

[†] "Projects" include awards and individual studies. "Mixed" means projects with both domestic and foreign components. "N/R" means that the data were not reported to the Commission. Departments/agencies that appear italicized reported that they were unable to provide complete data. See Appendix II for additional detail.

APPENDIX I: Scope and Volume of Federally Supported Human Subjects Research V

DEPARTMENT/ AGENCY	LOCATION	FY06	FY07	FY08	FY09	FY10
NSF	Domestic	1,820	2,271	2,626	2,984	3,049
	Foreign	0	0	1	4	2
	Mixed	0	0	0	0	0
	Unknown	0	0	0	0	0
SSA	Domestic	15	18	17	14	13
	Foreign	0	0	0	0	0
	Mixed	0	0	0	0	0
	Unknown	0	0	0	0	0
USAID	Domestic	26	28	20	15	14
	Foreign	126	121	82	44	45
	Mixed	7	8	4	1	3
	Unknown	0	0	0	4	0
USDA	Domestic	111	118	193	198	258
	Foreign	4	3	5	2	5
	Mixed	1	1	17	1	4
	Unknown	2	0	5	0	5
VA¶	Domestic	0	0	0	0	0
	Foreign	0	0	0	0	0
	Mixed	0	1	2	4	9
	Unknown	16,763	16,730	16,704	16,379	15,406
All	Domestic	29,383	32,280	31,741	33,819	36,046
	Foreign	721	924	984	888	931
	Mixed	666	872	1115	1391	1624
	Unknown	20,821	19,342	19,818	18,398	16,785
Total		51,591	53,418	53,658	54,496	55,386

† "Projects" include awards and individual studies. "Mixed" means projects with both domestic and foreign components. "N/R" means that the data were not reported to the Commission. Departments/agencies that appear italicized reported that they were unable to provide complete data. See Appendix II for additional detail.
¶ Although most VA research normally takes place in the United States, VA did not have data to assure the Commission that all of its research for which it did not specify a site country took place domestically.

Table I.8 Location of HHS Human Subjects Projects Over Time[†]

UNIT	LOCATION	FY06	FY07	FY08	FY09	FY10
AHRQ	Domestic	668	634	598	655	893
	Foreign	0	0	0	0	0
	Mixed	0	0	0	0	0
	Unknown	0	0	0	0	5
ASPR	Domestic	2	3	4	1	4
	Foreign	2	0	0	2	0
	Mixed	1	2	0	0	0
	Unknown	3	3	6	1	8
CDC	Domestic	1,637	1,264	1,182	1,078	1,077
	Foreign	85	104	177	235	220
	Mixed	16	17	21	21	20
	Unknown	1	2	0	0	0
CMS	Domestic	0	0	0	0	0
	Foreign	0	0	0	0	0
	Mixed	0	0	0	0	0
	Unknown	0	0	0	0	0
FDA	Domestic	85	102	121	191	200
	Foreign	2	1	1	1	0
	Mixed	0	0	0	0	0
	Unknown	3	0	0	0	0
HRSA	Domestic	45	50	50	43	35
	Foreign	0	0	0	0	0
	Mixed	2	1	0	0	0
	Unknown	0	0	0	0	0
IHS	Domestic	254	277	238	239	251
	Foreign	0	0	0	0	0
	Mixed	0	0	0	0	0
	Unknown	0	0	0	0	0

[†] "Projects" include awards and individual studies. "Mixed" means projects with both domestic and foreign components. See Appendix II for additional detail.

APPENDIX I: Scope and Volume of Federally Supported Human Subjects Research

UNIT	LOCATION	FY06	FY07	FY08	FY09	FY10
NIH	Domestic	21,473	20,827	20,088	21,409	21,033
	Foreign	281	296	296	265	261
	Mixed	635	834	1,064	1,354	1,557
	Unknown	7	1,253	1,261	1,000	1,040
NVPO	Domestic	0	0	0	0	0
	Foreign	0	0	0	0	0
	Mixed	0	0	0	0	0
	Unknown	61	14	39	0	0
OAH‡	Domestic	x	x	x	x	30
	Foreign	x	x	x	x	0
	Mixed	x	x	x	x	0
	Unknown	x	x	x	x	1
OPA	Domestic	12	16	22	17	16
	Foreign	0	0	0	0	0
	Mixed	0	0	0	0	0
	Unknown	0	0	0	0	0
SAMHSA	Domestic	0	0	0	0	0
	Foreign	0	0	0	0	0
	Mixed	0	0	0	0	0
	Unknown	0	0	0	0	0
All	Domestic	24,176	23,173	22,303	23,633	23,539
	Foreign	370	401	474	503	481
	Mixed	654	854	1,085	1,375	1,577
	Unknown	75	1,272	1,306	1,001	1,054
Total		25,275	25,700	25,168	26,512	26,651

† "Projects" include awards and individual studies. "Mixed" means projects with both domestic and foreign components. See Appendix II for additional detail.
‡ OAH was established in 2010.

Table I.9 Total Extramural Award Funding (Human Subjects Projects) Over Time (in $)[†]

DEPARTMENT/AGENCY	FY06	FY07	FY08	FY09[‡]	FY10[‡]
CIA[§]	N/R	N/R	N/R	N/R	N/R
CPSC	0	0	0	0	612,662
DHS[¶]	N/R	0	0	0	0
DOC	1,575,482	819,419	1,735,902	1,609,793	3,389,309
DOD	*162,285*	*0*	*0*	*0*	*0*
DOE	46,865,718	52,400,201	44,218,307	46,772,089	34,528,425
DOJ	51,306,495	57,291,788	33,820,452	63,808,343	113,365,616
DOT	5,268,756	20,410,017	3,304,967	14,373,087	4,125,572
ED	*151,777,538*	*611,483,875*	*705,112,958*	*802,093,966*	*1,120,977,802*
EPA	27,280,526	14,764,372	43,740,633	18,018,128	22,252,025
HHS	12,140,348,243	13,025,906,747	11,727,308,402	13,543,732,524	14,172,966,147
HUD	3,863,530	7,196,379	3,570,404	10,889,720	15,503,513
NASA	2,075,000	*3,100,000*	3,822,000	*4,858,000*	6,583,124
NSF	475,612,817	639,525,864	774,954,954	998,513,105	1,013,447,119
SSA	33,676,780	31,367,788	31,078,994	23,642,888	16,039,115
USAID	39,023,094	34,526,585	36,113,044	30,310,172	32,660,180
USDA	18,859,721	25,786,633	81,409,010	58,552,343	98,614,148
VA	0	0	0	0	0
Total	12,997,695,985	14,524,579,668	13,490,190,027	15,617,174,158	16,655,064,757

[†] The total, de-duplicated funding amount was calculated as described in the methods. "N/R" means that the data were not reported to the Commission. Departments/agencies that appear italicized reported that they were unable to provide complete extramural funding data. See Appendix II for additional detail.
[‡] FY09 and FY10 funding includes American Recovery and Reinvestment Act (ARRA) funding.
[§] The CIA did not submit project-level data to the Commission's database because these data are confidential (although not classified).
[¶] DHS reported that there "are no earlier data" than FY07.

APPENDIX I: Scope and Volume of Federally Supported Human Subjects Research

Table I.10 Total HHS Extramural Award Funding (Human Subjects Projects) Over Time (in $)[†]

UNIT	FY06	FY07	FY08	FY09[‡]	FY10[‡]
AHRQ	103,132,977	85,009,499	84,190,585	108,927,568	501,164,798
ASPR	*1,295,992,119*	*1,333,999,855*	*329,190,000*	*254,329,277*	*372,039,662*
CDC	*0*	*7,213,883*	*7,213,883*	*0*	*840,000*
CMS	0	0	0	0	0
FDA	*12,840,244*	*16,527,562*	*14,048,750*	*13,008,554*	*16,947,213*
HRSA	24,066,586	28,774,579	28,678,683	25,630,790	24,879,696
IHS	1,935,769	1,794,339	1,906,226	2,384,267	3,055,331
NIH	10,684,023,688	11,547,265,574	11,251,479,555	13,136,308,827	13,206,860,106
NVPO	15,994,885	2,275,174	6,320,016	0	0
OAH[§]	x	x	x	x	44,018,655
OPA	2,361,975	3,046,282	4,280,704	3,143,241	3,160,686
SAMHSA	0	0	0	0	0
Total	12,140,348,243	13,025,906,747	11,727,308,402	13,543,732,524	14,172,966,147

[†] The total, de-duplicated funding amount was calculated as described in the methods. Departments/agencies that appear italicized reported that they were unable to provide complete extramural funding data. See Appendix II for additional detail.
[‡] FY09 and FY10 funding includes American Recovery and Reinvestment Act (ARRA) funding.
[§] OAH was established in 2010.

Table I.11 Top 20 (of 117) Human Subjects Project Site Countries by Number of Projects, FY10†

SITE COUNTRY	USAID	CIA‡	CPSC	USDA	DOC	DOD	ED	DOE	HHS
United States	17	N/R	1	262	23	6,858	1,538	349	25,121
Canada				2		22	1	3	86
Peru	1					69	4	1	14
Kenya	8			1		10	3		53
Egypt						61	3		
South Africa	13						5	2	30
United Kingdom	2					14	3		28
Uganda	12					1			28
India	8						7		25
China	1					2	10	2	21
Thailand						12	1	1	16
Australia	1					8			15
Bangladesh	4			1					19
Brazil						3	3		16
Tanzania	4					1	5		8
Malawi	5			1					11
Indonesia	1					11	1		
France						5	3		4
Mexico	1			1		1	4		3
Sweden						6		2	4
Foreign									1,557
Blank or N/A¶				5		2	316		1,054

APPENDIX I: Scope and Volume of Federally Supported Human Subjects Research   V

DHS	HUD	DOJ	DOT	VA	EPA	NASA	NSF	SSA	TOTAL
10	18	215	56	9	44	133	3,049	13	37,716
			1		2				117
				1					90
									75
									64
					1				51
							1		48
									41
									40
									36
					1				31
									24
									24
				1			1		24
				1					19
									17
									13
									12
						2			12
									12
									1,557
				15,406§		2			16,785

† "Projects" include awards and individual studies. Departments/agencies that appear italicized reported that they were unable to provide complete extramural funding data. See Appendix II for additional detail
‡ "N/R" means that the data were not reported to the Commission. The CIA did not submit project-level data to the Commission's database because these data are confidential (although not classified), but the agency did advise the Commission that all of its human subjects research takes place in the United States.
§ Although most VA research normally takes place in the United States, VA did not have data to assure the Commission that all of its research for which it did not specify a site country took place domestically.
¶ "Blank" and "N/A" mean that the department/agency did not report a site country. See Appendix II for additional detail.

Table I.12 Top 20 (of 68) Extramural Awardee Institution Countries by Number of Human Subjects Projects, FY10[†]

INSTITUTION COUNTRY	TOTAL
United States	29,142
Canada	94
United Kingdom	34
Australia	19
China	16
South Africa	15
Bangladesh	14
Germany	11
India	11
New Zealand	11
France	10
Netherlands	10
Switzerland	10
Brazil	9
Kenya	9
Sweden	9
Israel	8
Peru	8
Uganda	7
Thailand	6
Invalid, N/A, or Unknown[‡]	607

[†] "Projects" include awards and individual studies. See Appendix II for additional detail.
[‡] "Invalid" means that text other than an institution name was entered in the "Institution name" column; "N/A" means that an institution name field was left blank; and "Unknown" means that the location of the awardee institution could not be determined with certainty, e.g., if an individual was listed as the awardee.

APPENDIX I: Scope and Volume of Federally Supported Human Subjects Research   V

Table I.13 Location of Unique Extramural (Human Subjects Projects) Awardee Institutions, FY10†

LOCATION	TOTAL
United States	2,867
Foreign	189
Invalid, N/A, or Unknown‡	60
Total	3,116

† "Projects" include awards and individual studies. See Appendix II for additional detail.
‡ "Invalid" means that text other than an institution name was entered in the "Institution name" column; "N/A" means that an institution name field was left blank; and "Unknown" means that the location of the awardee institution could not be determined with certainty, e.g., if an individual was listed as the awardee.

Table I.14 Proportion of Extramural Human Subjects Projects with Known Award Amounts†

DEPARTMENT/AGENCY	FY06	FY07	FY08	FY09	FY10
CIA‡	N/R	N/R	N/R	N/R	N/R
CPSC	—	—	—	—	1.000
DHS§	N/R	0.000	0.000	0.000	0.000
DOC	1.000	1.000	1.000	1.000	1.000
DOD	0.001	0.000	0.000	0.000	0.000
DOE	1.000	1.000	1.000	1.000	1.000
DOJ	1.000	1.000	1.000	1.000	1.000
DOT	1.000	1.000	1.000	1.000	1.000
ED	0.468	0.789	0.791	0.776	0.585
EPA	1.000	1.000	1.000	1.000	1.000
HHS	0.986	0.988	0.985	0.984	0.986
HUD	1.000	1.000	1.000	1.000	1.000
NASA	1.000	0.867	1.000	0.957	1.000
NSF	1.000	1.000	1.000	1.000	1.000
SSA	1.000	1.000	1.000	1.000	1.000
USAID	1.000	1.000	1.000	1.000	1.000
USDA	1.000	1.000	1.000	1.000	1.000
VA	—	—	—	—	—
Weighted Average	0.875	0.878	0.876	0.912	0.892

† "Extramural" includes projects conducted solely extramurally and projects with an extramural component; "Projects" include awards and individual studies; "Known" means "Total Award $" is not N/A, NULL, or an "invalid" $0, i.e., a $0 indicating that the Department/Agency cannot link funding and project data. "—" means that the proportion is undefined because the denominator is 0. "N/R" means that the data were not reported to the Commission. See Appendix II for additional detail.
‡ The CIA did not submit project-level data to the Commission's database because these data are confidential (although not classified).
§ DHS reported that there "are no earlier data" than FY07.

Table I.15 Proportion of Human Subjects Projects with Known Site Country[†]

DEPARTMENT/AGENCY	FY06	FY07	FY08	FY09	FY10
CIA[‡]	N/R	N/R	N/R	N/R	N/R
CPSC	1.000	1.000	1.000	1.000	1.000
DHS[§]	N/R	1.000	1.000	1.000	1.000
DOC	1.000	1.000	1.000	1.000	1.000
DOD	0.407	0.807	0.748	0.851	1.000
DOE	1.000	1.000	1.000	1.000	1.000
DOJ	0.983	1.000	1.000	1.000	1.000
DOT	1.000	1.000	1.000	1.000	1.000
ED	0.514	0.937	0.946	0.938	0.840
EPA	1.000	1.000	1.000	1.000	1.000
HHS	0.997	0.951	0.948	0.962	0.960
HUD	1.000	1.000	1.000	1.000	1.000
NASA	0.892	1.000	1.000	1.000	0.985
NSF	1.000	1.000	1.000	1.000	1.000
SSA	1.000	1.000	1.000	1.000	1.000
USAID	1.000	1.000	1.000	0.938	1.000
USDA	0.983	1.000	0.977	1.000	0.982
VA	0.000	0.000	0.000	0.000	0.001
Weighted Average	0.596	0.638	0.631	0.662	0.697

[†] "Projects" include awards and individual studies; "Known" means the entry in the "Site Country" field is not N/A, NULL, or blank. Note that a response of "Foreign" is considered "Known" in this report. "N/R" means that the data were not reported to the Commission. See Appendix II for additional detail.
[‡] The CIA did not submit project-level data to the Commission's database because these data are confidential (although not classified).
[§] DHS reported that there "are no earlier data" than FY07.

APPENDIX I: Scope and Volume of Federally Supported Human Subjects Research V

Table I.16 Proportion of HHS Human Subjects Projects with Known Site Country[†]

UNIT	FY06	FY07	FY08	FY09	FY10
AHRQ	1.000	1.000	1.000	1.000	0.994
ASPR	0.625	0.625	0.400	0.750	0.333
CDC	0.999	0.999	1.000	1.000	1.000
FDA	0.967	1.000	1.000	1.000	1.000
HRSA	1.000	1.000	1.000	1.000	1.000
IHS	1.000	1.000	1.000	1.000	1.000
NIH	1.000	0.946	0.944	0.958	0.956
NVPO	0.000	0.000	0.000	—	—
OAH[‡]	x	x	x	x	0.968
OPA	1.000	1.000	1.000	1.000	1.000
Weighted Average	0.997	0.951	0.948	0.962	0.960

[†] "Projects" include awards and individual studies; "Known" means the entry in the "Site Country" field is not N/A, NULL, or blank. Note that a response of "Foreign" is considered "Known" in this report. "—" means that the proportion is undefined because the denominator is 0. See Appendix II for additional detail.
[‡] OAH was established in 2010.

Table I.17 Proportion of Human Subjects Projects with Known Subject Count[†]

DEPARTMENT/AGENCY	FY06	FY07	FY08	FY09	FY10
CIA[‡]	N/R	N/R	N/R	N/R	N/R
CPSC	1.000	1.000	1.000	1.000	1.000
DHS[§]	N/R	0.000	0.000	0.000	0.000
DOC	0.857	0.846	0.853	0.852	0.870
DOD	0.000	0.001	0.005	0.005	0.005
DOE	0.867	0.905	0.908	0.914	0.950
DOJ	0.125	0.161	0.291	0.164	0.097
DOT	0.577	0.346	0.429	0.404	0.411
ED	0.000	0.000	0.000	0.000	0.000
EPA	0.167	0.220	0.256	0.143	0.000
HHS	0.073	0.075	0.083	0.084	0.085
HUD	0.000	0.000	0.000	0.000	0.000
NASA	0.108	0.144	0.215	0.203	0.787
NSF	0.000	0.000	0.000	0.000	0.000
SSA	1.000	1.000	1.000	1.000	1.000
USAID	0.126	0.121	0.179	0.375	0.403
USDA	0.525	0.492	0.350	0.398	0.706
VA	0.000	0.000	0.000	0.000	0.000
Weighted Average	0.046	0.046	0.050	0.051	0.055

[†] "Projects" include awards and individual studies; "Known" means the entry in the "# Participants" field is not N/A, NULL, or blank. "N/R" means that the data were not reported to the Commission. See Appendix II for additional detail.
[‡] The CIA did not submit project-level data to the Commission's database because these data are confidential (although not classified).
[§] DHS reported that there "are no earlier data" than FY07.

APPENDIX I: Scope and Volume of Federally Supported Human Subjects Research   V

Table I.18 Proportion of Human Subjects Projects with Known Exempt/Non-exempt Status[†]

DEPARTMENT/AGENCY	FY06	FY07	FY08	FY09	FY10
CIA[‡]	N/R	N/R	N/R	N/R	N/R
CPSC	0.000	0.000	0.000	0.000	0.000
DHS[§]	N/R	1.000	1.000	1.000	1.000
DOC	1.000	1.000	1.000	1.000	1.000
DOD	0.666	0.878	0.901	0.854	0.839
DOE	1.000	1.000	1.000	1.000	1.000
DOJ	0.467	0.432	0.512	0.414	0.472
DOT	0.385	0.269	0.200	0.277	0.268
ED	0.910	0.130	0.158	0.164	0.396
EPA	1.000	1.000	1.000	1.000	1.000
HHS	0.033	0.035	0.035	0.036	0.041
HUD	0.714	0.909	0.857	0.944	0.722
NASA	1.000	1.000	1.000	1.000	1.000
NSF	1.000	1.000	1.000	1.000	1.000
SSA	1.000	0.444	1.000	1.000	1.000
USAID	0.000	0.000	0.000	0.000	0.000
USDA	0.873	0.893	0.959	0.980	0.801
VA	0.000	0.000	0.000	0.000	0.000
Weighted Average	0.154	0.184	0.201	0.190	0.213

[†] "Projects" include awards and individual studies; "Known" means the entry in the "Exempt or Non-Exempt" field is not N/A, NULL, or blank. "N/R" means that the data were not reported to the Commission. See Appendix II for additional detail.
[‡] The CIA did not submit project-level data to the Commission's database because these data are confidential (although not classified).
[§] DHS reported that there "are no earlier data" than FY07.

Table I.19 Proportion of Human Subjects Projects with Known Number of Sites[†]

DEPARTMENT/AGENCY	FY06	FY07	FY08	FY09	FY10
CIA[‡]	N/R	N/R	N/R	N/R	N/R
CPSC	1.000	1.000	1.000	1.000	0.000
DHS[§]	N/R	0.000	0.000	0.000	0.000
DOC	1.000	0.923	0.941	0.926	1.000
DOD	0.000	0.012	0.008	0.028	0.006
DOE	0.998	0.997	0.997	0.997	1.000
DOJ	0.200	0.203	0.337	0.214	0.241
DOT	0.500	0.308	0.371	0.404	0.482
ED	0.000	0.000	0.000	0.000	0.000
EPA	0.472	0.512	0.907	0.429	0.217
HHS	0.001	0.001	0.000	0.000	0.001
HUD	0.000	0.000	0.000	0.000	0.000
NASA	0.892	1.000	0.909	0.872	0.985
NSF	0.000	0.000	0.000	0.000	0.000
SSA	0.933	0.944	1.000	1.000	1.000
USAID	0.547	0.688	0.755	0.875	0.403
USDA	0.864	0.877	0.832	0.846	0.816
VA	0.000	0.000	0.000	0.000	0.000
Weighted Average	0.015	0.017	0.017	0.018	0.017

[†] "Projects" include awards and individual studies; "Known" means the entry in the "# Sites [per country]" field is not N/A, NULL, or blank. "N/R" means that the data were not reported to the Commission. See Appendix II for additional detail.
[‡] The CIA did not submit project-level data to the Commission's database because these data are confidential (although not classified).
[§] DHS reported that there "are no earlier data" than FY07.

Appendix II: Human Subjects Research Landscape Project Methods

In order to respond to President Obama's charge, the Commission recognized that a critical first step would be to define and understand the landscape of "scientific studies supported by the Federal Government." Finding no comprehensive publicly available source for this information, the Commission asked the 18 federal departments and agencies that have adopted the Common Rule—and therefore were likely to support scientific studies with human subjects—to provide basic project-level data for department/agency-supported human subjects research in Fiscal Year 2006 to Fiscal Year 2010. These agencies are listed in Table I.1[1] and an overview of the Human Subjects Research Landscape Project is displayed in Figure II.1.

Commission Chair, Dr. Amy Gutmann, wrote to department/agencies regarding this request in early spring 2011. An example of the letter is provided in Figure II.2. As necessary, Commission staff clarified the data request with contacts at departments/agencies. The Commission asked departments/agencies to provide only data they maintained and that was readily available so that the Commission could respond to President Obama's charge in a timely manner. A summary of responsive data received is included in Table II.1.

Database and Electronic Data Collection Tools

The Commission engaged a contractor, SRA International, Inc. (SRA), to develop 1) an electronic data collection tool to assist departments/agencies in gathering data, 2) a website through which department/agencies could submit data (www.bioethics-rpd.net), and 3) a database in which to store these data, called the "Research Project Database" (RPD). The Commission, through SRA, also established a Help Desk to provide technical assistance to departments/agencies.

The Commission provided departments/agencies with the option to collect their data either in Microsoft Excel or XML format, and provided templates and instructions for each. (The data fields and instructions are listed in Table II.2.) These data collection tools were equipped with built-in data validations so that departments/agencies could pre-screen their data prior to upload to the RPD.

Registered department/agency users could access the password-protected RPD website to upload, delete, or review submitted data. Department/agency users uploaded data in a separate file for each fiscal year. The system validated all data fields upon upload, for example, to confirm that each "Study ID" (i.e., unique study identification number) was unique in a single fiscal year (and, therefore, that each study was listed only once per year). If data fields were found to have errors, the system provided the department/agency with an automated report explaining the errors encountered during the data validation along with a request to resubmit the data. If a department/agency did not enter "Site Data" (i.e., site country, number of sites per country, and number of participants per country) or "Other Federal Funding Data" (i.e., source of other federal funding and other federal funder identifier, such as an award number), the system displayed "warnings" asking the department/agency to either add these data, or to confirm that these data were not maintained or readily available. The department/agency could then add these data and resubmit, or confirm that these data were not maintained or readily available to bypass the warnings and submit the file as-is.

If a department/agency supported no human subjects research in a given fiscal year, Commission staff asked for written confirmation of that fact.[2]

Uploaded data were stored in an SRA-hosted SQL Server database. The original uploaded Excel and XML documents were also stored and retained on an SRA server. Following the data collection period, SRA exported the entire data set from the RPD into three "comma separated values" (.csv) files. The export process and naming conventions are detailed in Table II.3. Data were organized in three separate tables: (i) "study records" that provides project-level data; (ii) "site records" that captures Site Data; and (iii) "other federal funding records" that captures Other Federal Funding Data. A unique ID field common to all three tables allowed for linkage among them.

In the Human Subjects Research Landscape Project, the term "project" refers to a single line of data entered by a department/agency, whereas "study" refers to an individual human subjects research protocol or activity; and "award" refers to an extramural award, such as a grant or contract, which may fund more than one "study." The Commission defined "project" broadly in order to accommodate different department/agency record-keeping

APPENDIX II: Human Subjects Research Landscape Project Methods

systems. Although study-level data were preferred, some departments and agencies provided award-level data for extramural human subjects research. Additional definitions are listed in Table II.2.

Data Cleaning

Generally, if department/agency data passed the system's validations, the Commission accepted these submissions as-is. Nonetheless, minimal data cleaning was performed to facilitate analyses, which is detailed below.

In the SQL database, SRA performed one cleaning task:

- *Incorrect "unit" names.* Departments/agencies could specify individual "units" for data submission. For example, NASA submitted data for four units: Ames Research Center, Johnson Space Center, Kennedy Space Center, and Langley Research Center. In total, six submitted files incorrectly omitted a unit designation. SRA corrected these unit names in the SQL database.

Prior to initial analysis of the data, consultant statisticians, Norman P. Ross, M.S., Ph.D. and Philip Kalina, M.A., ran a number of checks on the data tables, including making sure that:

- All variables were in columns and observation records were in rows;
- There was one unique id for each project record; and
- All missing data had been identified and the appropriate code had been inserted in missing data cells.

Once the data were screened and checked, statisticians performed a comprehensive data cleaning process on the analytical database to remove anomalies that could be detected through statistical screening; for example, looking for missing values and contradictions within or between records, duplicates, and outliers. Before final analysis, the data were further cleaned as follows:

- *Projects removed from the analysis dataset.* Some departments/agencies noted in the "Other Comments" field that they were unable to delete or remove records from their data submissions. Based on a manual review of these comments, a small number of awards (eight) were moved out of the analysis dataset.

- *Addition of data submitted after close of the database.* HHS-ASPR supplemented its data submission after the database was closed. So that all data submitted to the Commission were accounted for in its analysis, these data were added to the SQL database and provided as a supplemental export to the statisticians to incorporate into the analysis dataset. In addition, one agency (Agricultural Research Service [USDA-ARS]) inadvertently uploaded the same FY06 file for two different units. When brought to its attention, USDA-ARS deleted the duplicate file and submitted corrected data after the database closed. These data were provided as a supplemental export to the consultant statisticians to incorporate into the analysis dataset. Finally, although DOD submitted aggregate data before the database closed, it submitted project-level data to the database after it was closed. These data also were provided as a supplemental export to the consultant statisticians to incorporate into the analysis set.

- *Units combined.* In the interest of simplifying and presenting data, some units were combined before analysis. The specific changes were:

  - Within USDA, all units starting with "ARS" were combined into one unit, Agricultural Research Service.

  - Within HHS, all units starting with "IHS" were combined into one unit, Indian Health Service.

  - Within HHS, all units starting with "National Institutes of Health" were combined into one unit, National Institutes of Health. NIH data were submitted in several parts due to limitations on the number of rows of data that could be entered into the Excel template. Because these divisions were arbitrary and not reflective of actual functional operating units, they were combined.

  - Within DOJ, all units starting with "OJP" were combined into one unit, Office of Justice Programs.

  - Within VA, all units were ignored. Like NIH, VA submitted its data in several parts due to limitations on the number of rows of data that could be entered into the Excel template. Because these divisions were arbitrary and not reflective of actual functional operating units, they were combined.

APPENDIX II: Human Subjects Research Landscape Project Methods

- Within DOD, all units were ignored under the same reasoning.
- *Study classification.* Where not apparent from department/agency data submissions, Commission staff asked for clarification about whether the submitted data were award level (i.e., each line of data corresponded an award) or study level (i.e., each line of data submitted corresponded to a single study). An additional column, "Study or Award Level," was added to the analysis dataset. Valid entries for this column were A (Award), S (Study), Q (Equivalent, where one award always supports a single study), U (Unclassifiable), and I (Intramural).
- *Site country data.* A few departments/agencies were able to state that all projects for which no country data were submitted took place in the United States.[3] These records were updated, and blanks were replaced with "United States."
- *Awardee institution names.* Departments/agencies submitted "Award Institution" names in a variety of formats (e.g., with differences in abbreviations, misspellings, etc.). Commission staff conducted a manual review of institution names and corrected obvious typographical errors and standardized institution names.
- *Awardee institution countries.* Awardee institution countries were manually added to the analysis dataset based on publicly available sources. If the awardee institution country could not be determined with certainty, the country was assigned a value of "Unknown." If a value other than an institution name was found in the "Award Institution" field (e.g., a department/agency mistakenly entered an abstract in this column), the country was assigned a value of "Invalid." If "N/A" had been entered in the institution name column, the country was assigned a value of "N/A."
- *Total extramural award amount.* "N/A" was not an accepted response in the "Total Award $ in FY" field. Because some department/agencies indicated that, although they entered "0" in the Total Award field, these data, in fact, were not available,[4] a new column, titled ExtraAwardFundVal0 (i.e., indicating whether "0" in the Total Award field was a "valid" 0 or an indication that data was not available) was added to the analysis dataset, where acceptable values were Y (Yes), N (No), and U (Unknown).

- *Extramural/intramural/both indication clarification.* NSF initially classified all projects as "both" in the "Intramural or Extramural" field (i.e., with both intramural and extramural components), but later clarified that all reported projects are extramural. This change was made accordingly in the analysis dataset.[5]

- *Duplicate awards.* Instances appeared in the database where, for a given fiscal year, a department/agency submitted lines of study-level data with identical award IDs and total award amounts. This is not necessarily indicative of an error, as a single award can fund multiple studies. Commission staff checked the affected awards in a publicly available database, USASpending.gov. If the award amount in the RPD matched the award amount in USASpending.gov, it was assumed that the award amount in the database did indeed reflect the total award amount and should not inadvertently be counted twice when making overall funding calculations. Because of these concerns, extramural funding tabulations were run in two ways: (i) by adding all total award amounts; and (ii) by adding all total award amounts except those identified as duplicates through the above process.

Following these cleaning processes, a final dataset was ready for analysis, tabulation, and statistical report generation.

Data Analysis

Following completeness and accuracy checks, the .csv files were read into a Microsoft Access database for analysis. The tables produced (included in Chapter 2 and Appendix I to this report) are based on descriptive tabulations and computations of relevant summary data. The descriptive summaries and tabulations presented provide a broad "landscape" view of the human subjects research activities being undertaken by participating departments/agencies both in the United States and in other countries. Tabulations were provided for Fiscal Year 2006 to Fiscal Year 2010 for all departments/agencies that provided data.

APPENDIX II: Human Subjects Research Landscape Project Methods

Empirical Advisory Group

The Commission convened the Empirical Advisory Group (EAG) to assist the Commission with its empirical work, comprised of two Commission members and six outside experts in bioethics, statistics, clinical trials, and qualitative research (listed in Table II.4). The EAG met on multiple occasions to discuss the Human Subjects Research Landscape Project and other empirical approaches that might be used to inform the Commission's response to the President's charge. The EAG advised the Commission concerning analysis and interpretation of the Human Subjects Research Landscape Project data.

Limitations

The Human Subjects Research Landscape Project provides information that characterizes human subjects research projects supported by the federal government. While these data are extensive, they must be interpreted with some limitations in mind. These limitations include:

The information was reported by departments/agencies and was not independently audited or verified. As such the completeness of reporting cannot be verified.[6]

Each department/agency determined what constituted "relevant" work, which may have contributed to reporting bias as well as difficulty in comparing data across departments/agencies. The Commission asked departments/agencies to report all human subjects research projects, but definitions of "human subjects research" can vary across departments/agencies.[7]

Not all departments/agencies provided all of the information requested. Accordingly, there may be distorted estimates of some summary statistics (e.g., total number of studies, funding/award information), further complicating making meaningful comparison within and between departments/agencies as well as comparisons over time. In addition, NIH provided two sets of intramural data retrieved from two different databases (IMPACII and Protrak), which have overlaps. Without looking through both sets of data individually, NIH could not be sure of the extent of the overlap or eliminate overlaps.[8] Thus, NIH intramural projects may thus be over-reported in these analyses.

A single extramural award can fund multiple studies. Thus, a department/agency's total extramural funding, calculated by summing relevant "Total Award $" fields, is likely an overestimate to the extent that the award funding reported may fund more projects than the single project listed.[9] Moreover, because some departments/agencies submitted award-level data and others submitted project-level data, the number of "projects" reported in the database is likely an underestimate of the total number of human subjects studies supported by the government because some projects may correspond to awards that fund more than one study.

The Human Subjects Research Landscape Project does not provide a robust understanding of research that was not reported because it is classified or because of national security concerns.[10]

APPENDIX II: Human Subjects Research Landscape Project Methods   V

Endnotes

1. Because HHS is the largest government supporter of human subjects research, the Human Subjects Research Landscape Project results are often presented in more detail for HHS Operating Divisions. Table I.1 also lists the HHS Operating Divisions that responded to the Commission's data request.

2. E-mail Correspondence: Theron Pride, DOJ, to Michelle Groman, PCSBI. (2011, August 2 and 2011, September 13); Phillip Smith, IHS, to Michelle Groman. (2011, September 20); Lori Putman, DOT, to Michelle Groman, PCSBI. (2011, October 17); Mark Grabowsky, NVPO, to Michelle Groman, PCSBI. (2011, October 3); Jeffery Rodamar, ED, to Michelle Groman, PCSBI. (2011, September 6); Richard Legault, DHS, to Michelle Groman, PCSBI. (2011, September 15); Mala Adiga, DOJ, to Michelle Groman, PCSBI. (2011, August 9; 2011, September 7, 2011; and 2011, September 12); MJ Fiocco, DOT, to Michelle Groman, PCSBI. (2011, August 26); Krista Fletcher, SAMHSA, to Michelle Groman, PCSBI. (2011, August 2); Amy Farb, OAH, to Michelle Groman, PCSBI. (2011, August 5); Memorandum from Jacquelyn White, CMS, to Dawn Smalls, CMS, Request from the Presidential Commission for the Study of Bioethical Issues for Information on Human Subjects Scientific Research OS#071220111044, July 26, 2011.

3. E-mail Correspondence: Valerie Bonham, PCSBI, to Kevin Neary, HUD. (2011, October 7); Alan Trachtenberg, IHS, to Michelle Groman, PCSBI. (2011, September 28); Barbara DeCausey, CDC, to Michelle Groman, PCSBI. (2011, October 14). Preeti Kanodia, HRSA, to Michelle Groman, PCSBI (2011, November 8). NIH explained that for awards to domestic institutions, it reported "United States" in the site country field; for awards to foreign institutions, it reported the name of the awardee country in the site country field; and for awards to domestic institutions that have a foreign component; it reported "United States" and "Foreign" in the site country field. Sarah Carr, NIH, to Michelle Groman, PCSBI. (2011, October 5). E-mail Correspondence. Thus, NIH projects understood as "foreign" (as opposed to "mixed," or with foreign and domestic components) represent direct awards to foreign institutions. Where a project reported no Site Data, a placeholder site record was created with blank values for country, sites, and participants. In addition, duplicate site records were removed from the analysis dataset; a small number of almost-exact duplicates were removed upon agency confirmation. Francis Chesley, AHRQ, to Michelle Groman, PCSBI. (2011, November 2). E-mail Correspondence.

4. Some departments/agencies and units indicated that they could not link some or all protocol-level data with extramural funding data. Letter from Richard Legault, DHS, to Valerie Bonham, PCSBI. (September 29, 2011). E-mail Correspondence: Michelle Groman, PCSBI, to Patty Decot, DOD. (2011, October 27); Jeffery Rodamar, ED, to Michelle Groman, PCSBI. (2011, September 6); Rhondalyn Cox, FDA, to Michelle Groman, PCSBI. (2011, August 5 and 2011, October 14); Barbara DeCausey, CDC, to Michelle Groman, PCSBI. (2011, October 13); Jeffrey Hill, NASA, PCSBI. (2011, August 2).

5. Michelle Groman, PCSBI, to Myron Gutmann, NSF. (2011, October 24). E-mail Correspondence.

6. For example, it cannot be stated with certainty that if the same project was reported in several fiscal years that the award amount, number of participants, etc., listed in each fiscal year data

set was the amount/number specific to that fiscal year or if the totals were repeated year after year. Similarly, for awards where ARRA funding was indicated, it is unclear whether the reported award amount is entirely or partially ARRA funded.

7   Other terms may be defined differently by different agencies as well, such as "extramural" and "intramural."

8   Sarah Carr, NIH, to Valerie Bonham, PCSBI. (2011, October 5). E-mail Correspondence.

9   Total extramural funding was calculated by summing relevant "Total Award $" fields rather than relevant "Total Extramural Study $" fields because the mean response rate for this latter variable was less than 17 percent.

10 For example, the CIA did not submit project-level data to the RPD because "the application by the C.I.A. of certain research results may implicate intelligence sources and methods, and thus cannot be discussed in the public domain." Letter from V. Sue Bromley, Associate Deputy Director, Central Intelligence Agency to Amy Gutmann, Ph.D., Chair, Presidential Commission for the Study of Bioethical Issues. (November 15, 2011). The CIA confirmed that all CIA-sponsored human subjects research is conducted in the United States – not abroad. CIA personnel also met with Commission staff to discuss the CIA's human subjects research portfolio and made records available to appropriately cleared Commission staff. In addition, the Department of Energy provided de-identified data about three human terrain mapping projects that have not been accounted for in the RPD.

APPENDIX II: Human Subjects Research Landscape Project Methods

V

Figure II.1 Human Subjects Research Landscape Project Overview

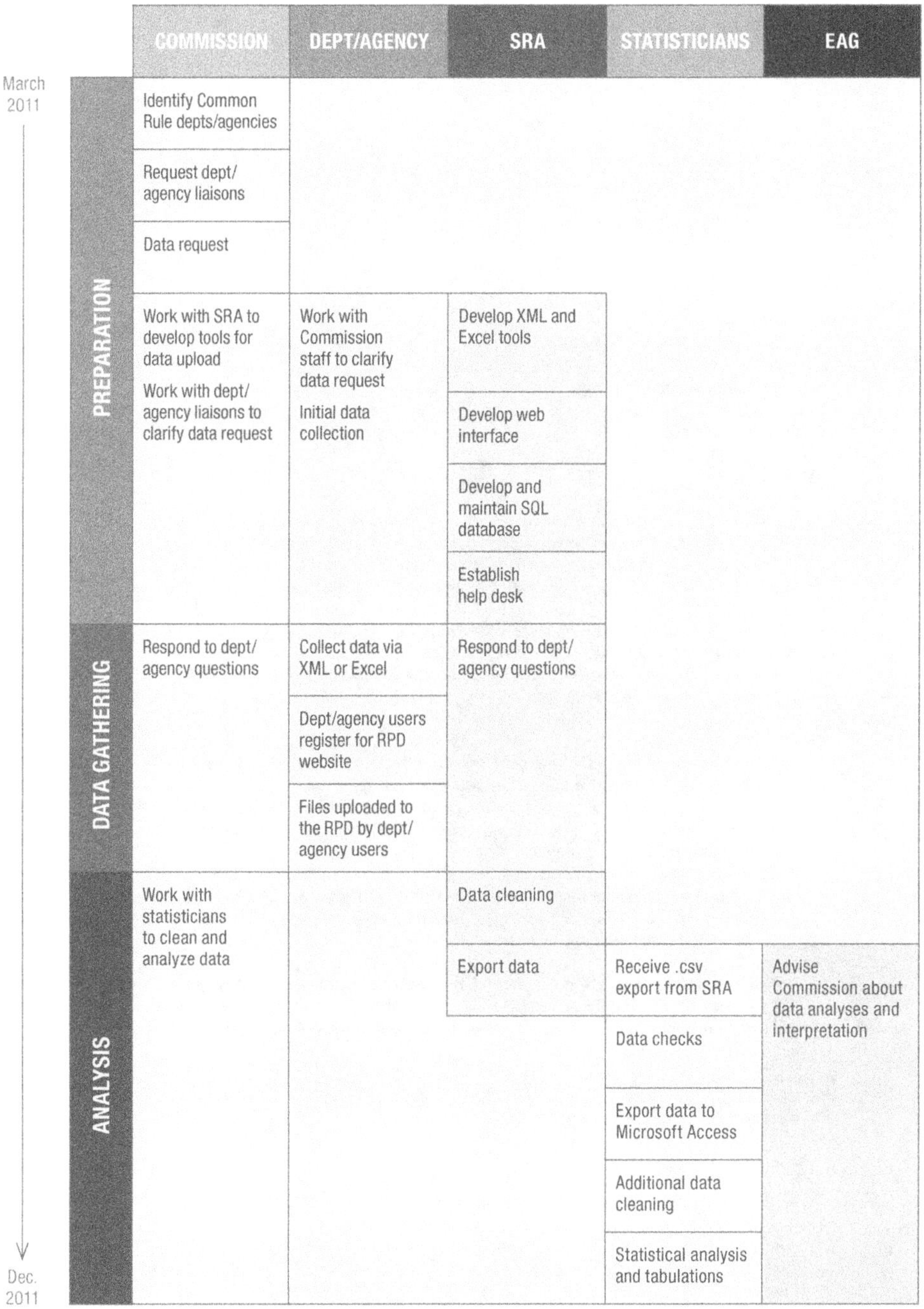

	COMMISSION	DEPT/AGENCY	SRA	STATISTICIANS	EAG
PREPARATION (March 2011)	Identify Common Rule depts/agencies				
	Request dept/agency liaisons				
	Data request				
	Work with SRA to develop tools for data upload	Work with Commission staff to clarify data request	Develop XML and Excel tools		
	Work with dept/agency liaisons to clarify data request	Initial data collection	Develop web interface		
			Develop and maintain SQL database		
			Establish help desk		
DATA GATHERING	Respond to dept/agency questions	Collect data via XML or Excel	Respond to dept/agency questions		
		Dept/agency users register for RPD website			
		Files uploaded to the RPD by dept/agency users			
ANALYSIS	Work with statisticians to clean and analyze data		Data cleaning		
			Export data	Receive .csv export from SRA	Advise Commission about data analyses and interpretation
				Data checks	
				Export data to Microsoft Access	
				Additional data cleaning	
				Statistical analysis and tabulations	

(Dec. 2011)

173

Figure II.2 Sample Letter from Dr. Amy Gutmann to Department/Agency Liaison

PRESIDENTIAL COMMISSION FOR THE STUDY OF BIOETHICAL ISSUES

March 25, 2011

Dr. Warren Lux
Human Subjects Research Review Official
Director of the Program in Human Research Ethics
Environmental Protection Agency
1200 Pennsylvania Ave., NW
Washington, DC 20460

Dear Dr. Lux:

As you know, President Obama charged the Presidential Commission for the Study of Bioethical Issues (the "Commission") to conduct a "thorough review of human subjects protection to determine if Federal regulations and international standards adequately guard the health and well-being of participants in scientific studies supported by the Federal Government."[1] Towards this end, the Commission is seeking comprehensive and accurate data about the volume of scientific studies supported by the Environmental Protection Agency (the "Agency") that involve human participants and information about the standards for protection of human participants, both domestically and internationally. Specifically, the Commission is asking the Agency to:

1. Provide nature, volume, and spending data (meaning dollar amounts and number of studies, sites, and participants) for scientific studies supported by the Agency that involve human subjects for fiscal years 2006-2010, as well as any related additional volume and/or trend data that the Agency may keep. Please provide data for studies occurring domestically and studies occurring internationally, and distinguish data by country. The Commission staff is developing an electronic means through which to collect these data in a uniform and efficient manner, and will contact you regarding this system as soon as it is available. Enclosed, please find a table detailing the data fields that will be requested in the electronic system.

2. Identify the Agency's regulations that guard the health and well-being of human participants in scientific studies supported by the Agency, and any additional guidances, policies, summaries, or explanations of the same that the Agency may distribute.

---

[1] The complete charge is available at http://www.whitehouse.gov/the-press-office/2010/11/24/presidential-memorandum-review-human-subjects-protection.

1425 NEW YORK AVENUE, NW, SUITE C-100, WASHINGTON, DC 20005
PHONE 202-233-3960   FAX 202-233-3990   WWW.BIOETHICS.GOV

APPENDIX II: Human Subjects Research Landscape Project Methods

Dr. Warren Lux
March 25, 2011
Page -2-

3. Identify the international standards with which the Agency complies or that it considers in guarding the health and well-being of human participants in scientific studies supported by the Agency, and any additional guidances, policies, summaries, or explanations of the same that the Agency may distribute.

The Commission is to deliver its final report later this year, and is, therefore, working under a very tight deadline. The Commission would very much appreciate if you could supply the above-requested information by April 27, 2011. The Commission staff would be happy to accept information on a rolling basis as soon as it is available and work with you to facilitate these requests.

A member of the Commission staff will contact you to follow up on these requests. The Commission staff will also contact you with any further requests for information if they arise. In the meantime, please do not hesitate to contact Ms. Valerie Bonham, the Commission's Executive Director, at (202) 233-3962 or Valerie.Bonham@bioethics.gov if you have any questions.

In addition, as described in the enclosed Federal Register notice, the Commission has requested public comment on the Federal and international standards for protecting the health and well-being of participants in scientific studies supported by the Federal Government. The Commission welcomes the Agency to submit comments and information in response to this request for public comment as well.

Thank you in advance for your help and consideration.

Sincerely,

Amy Gutmann, Ph.D.
Chair

Enclosures

Table II.1 Responsive Data Received[†]

DEPARTMENT/ AGENCY	UNIT	DATA SUBMITTED TO RPD				
		FY06	FY07	FY08	FY09	FY10
Agency for International Development		Y	Y	Y	Y	Y
Central Intelligence Agency		N	N	N	N	N
Consumer Product Safety Commission		Y	Y	Y	Y	Y
Department of Agriculture	Agricultural Research Service	Y	Y	Y	Y	Y
	Economic Research Service	Y	Y	Y	Y	Y
	National Institute of Food and Agriculture	Y	Y	Y	Y	Y
Department of Commerce		Y	Y	Y	Y	Y
Department of Defense		Y	Y	Y	Y	Y
Department of Education	Institute for Educational Sciences	Y	Y	Y	Y	Y
	Office for English Language Education	N-None	N-None	N-None	N-None	N-None
	Office for Elementary and Secondary Education	N	N	N	N	Y
	Office for Innovation and Improvement	N	N	N	N	Y
	Office for Postsecondary Education (including Fulbright-Hays fellowships)	Y	Y	Y	Y	Y
	Office of Planning, Evaluation & Policy Development	N	N	N	N	Y
	Office of Safe and Drug Free Schools	N	N	N	N	Y
	Office for Special Education and Rehabilitative Services (including National Institute for Disability and Rehabilitation Research)	N	Y	Y	Y	Y
	Office for Vocational and Adult Education	N	N	N	N	N
Department of Energy		Y	Y	Y	Y	Y

[†] "Y" indicates that the department/agency or unit submitted data to the RPD for the given fiscal year. "N-None" indicates that the department/agency or unit informed the Commission that it did not support human subjects research in the given fiscal year. "N" indicates that the department/agency or unit did not submit data to the RPD for the given fiscal year. The CIA did not submit project-level data to the Commission's database because these data are confidential (although not classified). Letter from V. Sue Bromley, Associate Deputy Director, CIA to Amy Gutmann, Ph.D., Chair, PCSBI. (November 15, 2011). ED did not upload data as summarized here, but also reported that "OESE, OII, OPEPD and OVAE have very few studies that fall under the Common Rule." Jeffery Rodamar, ED, to Michelle Groman, PCSBI. (2011, September 14). E-mail Correspondence. DHS reported that it had "no earlier data" than FY07. Richard Legault, DHS, to Michelle Groman, PCSBI. (2011, September 15). E-mail Correspondence.

APPENDIX II: Human Subjects Research Landscape Project Methods

DEPARTMENT/ AGENCY	UNIT	DATA SUBMITTED TO RPD				
		FY06	FY07	FY08	FY09	FY10
Department of Health and Human Services	Agency for Healthcare Research and Quality	Y	Y	Y	Y	Y
	Assistant Secretary for Preparedness and Response	Y	Y	Y	Y	Y
	Centers for Disease Control and Prevention	Y	Y	Y	Y	Y
	Centers for Medicare and Medicaid Services	N-None	N-None	N-None	N-None	N-None
	Food and Drug Administration	Y	Y	Y	Y	Y
	Health Resources and Services Administration	Y	Y	Y	Y	Y
	Indian Health Service[‡]	Y	Y	Y	Y	Y
	National Institutes of Health	Y	Y	Y	Y	Y
	OASH National Vaccine Program Office	Y	Y	Y	N-None	N-None
	Office of Adolescent Health[§]	N-None	N-None	N-None	N-None	Y
	Office of Population Affairs	Y	Y	Y	Y	Y
	Substance Abuse & Mental Health Services Administration	N-None	N-None	N-None	N-None	N-None
Department of Homeland Security		N	Y	Y	Y	Y
Department of Housing and Urban Development	Office of Healthy Homes & Lead Hazard Control	Y	Y	Y	Y	Y
	Office of Policy Development and Research	Y	Y	Y	Y	Y
Department of Justice	Bureau of Prisons	Y	Y	Y	Y	Y
	Federal Bureau of Investigation	Y	Y	Y	Y	Y
	Office of Community Oriented Policing Services	N-None	N-None	N-None	N-None	N-None
	Office of Justice Programs[¶]	Y	Y	Y	Y	Y
	Office on Violence Against Women	N-None	N-None	N-None	N-None	N-None

[‡] Within IHS, the Billings Area Office did not support human subjects research in FY09.
[§] Because it is a "new" office, OAH did not have FY06-FY09 data to report. Amy Farb, OAH, to Michelle Groman, PCSBI. (2011, August 5). E-mail Correspondence. OAH. About the Office of Adolescent Health. Retrieved from http://www.hhs.gov/ash/oah/about-us/ (last accessed December 8, 2011) ("OAH was established through the Consolidated Appropriations Act of 2010, within the Office of the Assistant Secretary for Health.").
[¶] Within OJP, the Bureau of Justice Assistance did not support human subjects research in FY07, FY08, or FY10 and the Office of Victims of Crime did not support human subjects research in FY06.

continued

Table II.1 Responsive Data Received[†]

DEPARTMENT/ AGENCY	UNIT	DATA SUBMITTED TO RPD				
		FY06	FY07	FY08	FY09	FY10
Department of Transportation	Federal Aviation Administration	Y	Y	Y	Y	Y
	Federal Highway Administration	Y	Y	Y	Y	Y
	Federal Motor Carrier Safety Administration	N-None	Y	Y	Y	Y
	Federal Railroad Administration	Y	Y	Y	Y	Y
	Maritime Administration	N-None	N-None	N-None	N-None	N-None
	National Highway Traffic Safety Administration[††]	Y	Y	Y	Y	Y
	Research and Innovative Technology Administration	Y	N-None	Y	Y	Y
Department of Veterans Affairs		Y	Y	Y	Y	Y
Environmental Protection Agency		Y	Y	Y	Y	Y
National Aeronautics and Space Administration	Ames Research Center	Y	Y	Y	Y	Y
	Johnson Space Center	Y	Y	Y	Y	Y
	Kennedy Space Center	Y	Y	Y	Y	Y
	Langley Research Center	Y	Y	Y	Y	Y
National Science Foundation		Y	Y	Y	Y	Y
Social Security Administration		Y	Y	Y	Y	Y

[†] "Y" indicates that the department/agency or unit submitted data to the RPD for the given fiscal year. "N-None" indicates that the department/agency or unit informed the Commission that it did not support human subjects research in the given fiscal year. "N" indicates that the department/agency or unit did not submit data to the RPD for the given fiscal year. The CIA did not submit project-level data to the Commission's database because these data are confidential (although not classified). Letter from V. Sue Bromley, Associate Deputy Director, CIA to Amy Gutmann, Ph.D., Chair, PCSBI. (November 15, 2011). ED did not upload data as summarized here, but also reported that "OESE, OII, OPEPD and OVAE have very few studies that fall under the Common Rule." Jeffery Rodamar, ED, to Michelle Groman, PCSBI. (2011, September 14). E-mail Correspondence. DHS reported that it had "no earlier data" than FY07. Richard Legault, DHS, to Michelle Groman, PCSBI. (2011, September 15). E-mail Correspondence.

[††] NHTSA data for FY06-FY09 does not include information about safety-related studies involving human subjects. Lori Putnam, DOT, to Michelle Groman, PCSBI. (2011, December 1). E-mail Correspondence.

APPENDIX II: Human Subjects Research Landscape Project Methods

Table II.2 Data Fields and Instructions

FIELD	INSTRUCTIONS
1. Study ID#	Enter a unique study identification number such as IRB or institute protocol number, IND number, or other unique identifier assigned by the Department/Agency. Note that award number is acceptable here but, because it is requested separately, an alternative identifier is preferred. NCT number is also acceptable here but, if available, should be provided in the "NCT#" field as well.
2. NCT# [N/A is option]	Enter NCT number, if available. For trials entered in ClinicalTrials.gov, ClinicalTrials.gov assigns a unique NCT identifier of the form NCTxxxxxxxx where each x is a numeric digit. Enter N/A if the Department/Agency does not maintain this data, or it is not readily available.
3. Title of Study	Enter the title of the study, as maintained by the Department/Agency. "Title of Study" is intended to be as specific as possible, with protocol title preferred. Award title may be substituted for protocol title when necessary. It is understood that an award may support more than one protocol.
4. Abstract [N/A is option]	Enter the study or award abstract if it is readily available. Enter N/A if the Department/Agency does not maintain this data, or it is not readily available.
5. PI(s)	Enter the name or names of the study's principal investigator(s). Names may be provided in any format, and can be separated by a "," or ";".
6. Year X of Y [N/A is option]	Enter the duration of the study, for example, "Year 2 of 4." "X" should be entered in reference to the fiscal year for which the Department/Agency is reporting. That means, for example, that a study reported in FY06 as "Year 2 of 4," would be reported in FY07 as "Year 3 of 4." Enter N/A if the Department/Agency does not maintain this data, or it is not readily available. If the full duration of the study is unknown, enter N/A for "Y."
7. Exempt or Non-Exempt [Ex/N] [N/A is option]	Enter Ex if the study is human subjects research "exempt" from 45 CFR 46 or applicable agency regulations. Enter N if the study is non-"exempt." Enter N/A if the Department/Agency does not maintain this data, or it is not readily available.
8. Total # Sites [N/A is option]	Enter the total number of locations where the study is being conducted, which may not correspond to where the approving IRB is located. Enter N/A if the Department/Agency does not maintain this data, or it is not readily available.
9. Site Country [N/A is option]	Enter all countries in which the study is being conducted. Enter each country in a separate row. Enter N/A if the Department/Agency does not maintain this data, or it is not readily available.
10. # Sites [Per Country] [N/A is option]	Enter the total number of locations where the study is being conducted in the listed country. Enter N/A if the Department/Agency does not maintain this data, or it is not readily available.
11. # Participants [Per Country] [N/A is option]	Enter the total number of participants in the listed country in the relevant fiscal year. There is no need to list participants for each site within the country separately. Enter N/A if the Department/Agency does not maintain this data, or it is not readily available.
12. ARRA Funded by Reporting Entity? [Y/N]	Enter Y if Department/Agency funding (if any) for the study is from American Recovery and Reinvestment Act (ARRA) funds. Enter N if Department/Agency funding (if any) for the study is not from American Recovery and Reinvestment Act (ARRA) funds.
13. Other Fed Funding? [Y/N] [N/A is option]	Enter Y if this study was funded in the reported fiscal year by another federal funder, in whole or in part. Enter N if this study was not funded in the reported fiscal year by another federal funder, in whole or in part. Enter N/A if the Department/Agency does not maintain this data, or it is not readily available.
14. Source of Other Fed Funding?	If this study was funded in the reported fiscal year by another federal funder, in whole or in part, select the Department/Agency that is the source of that federal funding. If more than one, enter each Department/Agency that is the source of other federal funding in a separate row. Select "Other" if the Department/Agency that is the source of other federal funding is not listed in the drop-down menu and, if known, enter its name in the "Other Comments" field. If this study was not funded in the reported fiscal year by another federal funder, in whole or in part, leave blank.

Table II.2 Data Fields and Instructions

FIELD	INSTRUCTIONS
15. Other Fed Funder Identifier [N/A is option]	If this study was funded in the reported fiscal year by another federal funder, in whole or in part, enter a study identification number assigned to the study by the Department/Agency that is the source of other federal funding, such as award number or IRB protocol number, if known and readily available. Enter N/A if not known or readily available. If this study was not funded in the reported fiscal year by another federal funder, in whole or in part, leave blank.
16. Other Non-Fed Funding? [Y/N] [N/A is option]	Enter Y if this study was funded in the reported fiscal year by another non-federal funder, in whole or in part. A non-federal funder could be, for example: foreign, state, or local governments or university, industry, non-profit, or philanthropic organizations. Enter N if this study was not funded in the reported fiscal year by another non-federal funder, in whole or in part. Enter N/A if the Department/Agency does not maintain this data, or it is not readily available.
17. Intramural or Extramural [I/E/B]	Enter I if the study is considered intramural by the Department/Agency. Enter E if the study is considered extramural by the Department/Agency. "Intramural," generally, means internal agency research programs. "Extramural," generally, means research supported by the Department/Agency through grant, cooperative agreement, contract, interagency agreement of any type, and "other transaction authority," e.g., 10 U.S.C. 2371 (DOD). For studies funded with both intramural and extramural monies, enter B.
18. Total Intramural Study $ in FY from Reporting Entity [N/A is option]	If intramural, enter the Department/Agency's intramural funding of the study in the reported fiscal year. Do not include funding from other federal or non-federal sources. This may be "0." If the Department/Agency does not track total study funding by project, aggregate amounts by fiscal year are acceptable, e.g., laboratory or program. Please provide an explanation in the "Other Comments" field. Enter N/A if the Department/Agency does not maintain this data, or it is not readily available. If extramural, leave blank.
19. Award ID#	If extramural, enter unique identification number assigned to the award by the Department/Agency. "Award" means grant, cooperative agreement, contract, interagency agreement of any type, and "other transaction authority," e.g., 10 U.S.C. 2371 (DOD). If intramural, leave blank.
20. Award Institution	If extramural, enter the name of the institution receiving the award. If intramural, leave blank.
21. Award Title	If extramural, enter the title of the award, as maintained by the Department/Agency. If intramural, leave blank.
22. Total Award $ in FY	If extramural, enter the amount of award extramural funding in the reported fiscal year. If intramural, leave blank.
23. Total Extramural Study $ in FY from Reporting Entity [N/A is option]	If extramural, enter the Department/Agency's extramural funding of the study in the reported fiscal year. Do not include funding from other federal or non-federal sources. This may be "0." Enter N/A if the Department/Agency does not maintain this data, or it is not readily available. If intramural, leave blank.
24. Direct Award $ in FY [N/A is option]	If extramural and "Total Study $" is not given, enter the amount of direct award funding in the reported fiscal year. Enter N/A if the Department/Agency does not maintain this data, or it is not readily available. If "Total Study $" is given, leave blank. If intramural, leave blank.
25. Indirect Award $ in FY [N/A is option]	If extramural and "Total Study $" is not given, enter the amount of indirect award funding in the reported fiscal year. Enter N/A if the Department/Agency does not maintain this data, or it is not readily available. If "Total Study $" is given, leave blank. If intramural, leave blank.
26. Other Comments	Enter any necessary explanations, as well as any additional information that may be helpful to the Commission about the listed study. If no other comments, leave blank.

APPENDIX II: Human Subjects Research Landscape Project Methods

Table II.3 SRA Methodology

Values as Stored in Database

SPREADSHEET FIELD	DATABASE FIELD	VALUE AS STORED IN DATABASE	VALUE FOR ANALYSIS (INVERSE OF VALUE STORED)
Study ID#	StudyID	As supplied	As supplied
NCT #	NCT	As supplied	As supplied
Title of Study	Title	As supplied	As supplied
Abstract	Abstract	As supplied	As supplied
PI(s)	PI	As supplied	As supplied
Year X	Year_X	Integer => As Supplied; N/A => NULL	Integer => As Supplied; NULL => N/A
Year Y	Year_Y	Integer => As Supplied; N/A => NULL	Integer => As Supplied; NULL => N/A
Exempt or Non-Exempt	Exempt	Ex => 1; N => 0; N/A =>NULL	1 => Ex; 0 => N; NULL =>N/A
Total # of Sites	Sites	Integer => As Supplied; N/A => NULL	Integer => As Supplied; NULL => N/A
ARRA Funded by Reporting	Arra	Y => 1; N => 0	1 => Y; 0 => N
Other Fed Funding	Other_Fed_Funding	Y => 1; N => 0; N/A => NULL	1 => Y; 0 => N; NULL => N/A
Other Non-Fed Funding	Other_NonFed_Funding	Y => 1; N => 0; N/A => NULL	1 => Y; 0 => N; NULL => N/A
Intramural or Extramural	Funding_Type	As Supplied	As Supplied
Total Intramural Study $ in FY	Intramural_Funding	IF FUNDING TYPE == E => NULL IF FUNDING TYPE == I,B {Money Value => As Supplied, without dollar formatting; N/A => NULL}	IF FUNDING TYPE == E => NULL IF FUNDING TYPE == I,B {Money Value => As Supplied; NULL => N/A}
Award ID#	Award_ID	IF FUNDING TYPE == I => NULL IF FUNDING TYPE == E,B => As Supplied	IF FUNDING TYPE == I => NULL IF FUNDING TYPE == E,B => As Supplied
Award Institution	Award_Inst	IF FUNDING TYPE == I => NULL IF FUNDING TYPE == E,B => As Supplied	IF FUNDING TYPE == I => NULL IF FUNDING TYPE == E,B => As Supplied

continued

Table II.3 SRA Methodology

Values as Stored in Database

SPREADSHEET FIELD	DATABASE FIELD	VALUE AS STORED IN DATABASE	VALUE FOR ANALYSIS (INVERSE OF VALUE STORED)
Award Title	Award_Title	IF FUNDING TYPE == I => NULL IF FUNDING TYPE == E,B => As Supplied	IF FUNDING TYPE == I => NULL IF FUNDING TYPE == E,B => As Supplied
Total Award $ in FY	Total_Funding	IF FUNDING TYPE == I => NULL IF FUNDING TYPE == E,B {Money Value => As Supplied, without dollar formatting; N/A => NULL}	IF FUNDING TYPE == I => NULL IF FUNDING TYPE == E,B {Money Value => As Supplied; NULL => N/A}
Total Extramural Study $ in FY from Reporting Entity	Extramural_Funding	IF FUNDING TYPE == I => NULL IF FUNDING TYPE == E,B {Money Value => As Supplied, without dollar formatting; N/A => NULL}	IF FUNDING TYPE == I => NULL IF FUNDING TYPE == E,B {Money Value => As Supplied; NULL => N/A}
Direct Award $ in FY	Direct_Funding	IF FUNDING TYPE == I => NULL IF FUNDING TYPE == E,B and EXTRAMURAL_FUNDING == N/A => { Money Value => As Supplied, without dollar formatting; N/A => NULL}	IF FUNDING TYPE == I => NULL IF FUNDING TYPE == E,B and EXTRAMURAL_FUNDING == N/A => {Money Value => As Supplied; NULL => N/A}
Indirect $ in FY	Indirect_Funding	IF FUNDING TYPE == I => NULL IF FUNDING TYPE == E,B and EXTRAMURAL_FUNDING == N/A => {Money Value => As Supplied, without dollar formatting; N/A => NULL}	IF FUNDING TYPE == I => NULL IF FUNDING TYPE == E,B and EXTRAMURAL_FUNDING == N/A => {Money Value => As Supplied; N/A => NULL}
Other Comments	Comments	As Supplied	As Supplied

APPENDIX II: Human Subjects Research Landscape Project Methods

Table II.4 Empirical Advisory Group

Robert M. Califf, MD
Vice Chancellor for Clinical Research
Duke University Medical Center
Director, Duke Translational
Medicine Institute

Ruth Faden, Ph.D., M.P.H.
Philip Franklin Wagley Professor
of Biomedical Ethics
Director, Johns Hopkins Berman
Institute of Bioethics
Professor, Department of Health Policy
and Management
Johns Hopkins Bloomberg
School of Public Health
Professor, Department of Medicine
Johns Hopkins School of Medicine

Kenneth A. Getz, M.B.A.
Founder and Board Chair
The Center for Information and Study
on Clinical Research Participation
Senior Research Fellow
Tufts Center for The Study of
Drug Development
Tufts University Medical School

Christine Grady, R.N., Ph.D.*
Acting Chief of the Department of Bioethics
National Institutes of Health
Clinical Center

Philip W. Lavori, Ph.D.
Professor, Health Research and Policy
Stanford School of Medicine

Bernard Lo, M.D.
Professor of Medicine
Director, Program in Medical Ethics
University of California San Francisco
National Program Director, Greenwall
Faculty Scholars Program in Bioethics

Kathleen M. MacQueen, Ph.D., M.P.H.
Senior Scientist, Behavioral &
Social Sciences
Coordinator of Interdisciplinary
Research Ethics
FHI 360

Daniel P. Sulmasy, M.D., Ph.D.*
Kilbride-Clinton Professor of Medicine
and Ethics, Department of Medicine and
Divinity School
Associate Director, The MacLean Center
for Clinical Medical Ethics
University of Chicago

* Commission member

Appendix III: U.S. Treatment/Compensation for Treatment Methods

Department of Defense Title 32: National Defense, Part 108 - Health Care Eligibility Under the Secretarial Designee Program and Related Special Authorities (2010) DOD Instruction Number 6000.08 (2007)	"Because subjects may be injured while participating in CI research, for research involving more than minimal risk as determined by the IRB having jurisdiction over the study, include in every intramural (in-house) CIP protocol an arrangement for treatment of any research-related injuries. Such arrangement in the protocol may be that all subjects are eligible DoD healthcare beneficiaries, that they are granted Secretarial designation as DoD healthcare beneficiaries under applicable Service regulations, or that specific obligations for such treatment have otherwise been undertaken." Section 6.2.4 (2007).
Department of Veterans Affairs 38 CFR §17.85 (2002)	"VA medical facilities shall provide necessary medical treatment to a research subject injured as a result of participation in a research project approved by a VA Research and Development Committee and conducted under the supervision of one or more VA employees." Part 17 Sec. 17.85(a).
Medicare Program Memorandum Intermediaries/Carriers: Claims Processing Instructions for Carriers, DMERCS, Intermediaries and Regional Home Health Intermediaries (RHHIs) for Claims Submitted for Medicare Beneficiaries Participating in Medicare Qualifying Trials (2000)	"This NCD states that Medicare covers: ... 2) reasonable and necessary items and services used to diagnose and treat complications arising from participation in *all* clinical trials." Background. "Routine costs DO include (and are therefore covered): ... Items and services that are medically necessary for the diagnosis or treatment of complications arising from the provision of an investigational item or service." Clinical Trial Services that Qualify for Coverage Section 2.
National Aeronautics and Space Administration NASA Procedural Requirements 7100.1 Protection of Human Research Subjects (Revalidated 7/7/08)	"The PI may ensure that the subject or the subject's beneficiaries receive compensation by means of insurance, worker's compensation, or the like in the event that the subject suffers illness, disease, injury, loss, or death as a direct result of the research. The lack of this provision <u>may</u> serve as a basis for disapproval of the research. Such provisions for compensation shall be required for all studies performed at a NASA Center, uses NASA equipment or facilities, or for which a NASA employee or on-site contractor is the principal investigator." Section 9.1.4.
National Institutes of Health Sheet 6 - Guidelines for Writing Informed Consent Documents (2006)	The Clinical Center of the NIH will provide short-term medical care for any injury resulting from your participation in research here. ... In general, no long-term medical care or financial compensation for research-related injuries will be provided by the National Institutes of Health, the Clinical Center, or the Federal Government. Section 4(s).
University of California Los Angeles Guidance and Procedure: Treatment and Compensation for Research Related Injury (last updated July 26, 2011)	"The University of California will provide to any injured subject any and all medical treatment reasonably necessary for any injury or illness which a human subject suffers as a direct result of participation in an authorized University activity covered by University policy on the protection of human subjects in research or reimburse the subject for the costs of such treatment, except when the injury or illness is a consequence of a medical research procedure which is designed to benefit the subject directly. University of California Policy on Treatment and Compensation for Injury in Research 1.

Appendix III: U.S. Treatment/Compensation for Treatment Methods V

University of Washington Human Subjects Manual, Section VII	"The University's policy on compensation for adverse events to human subjects is intended primarily to provide necessary medical care to subjects who sustain bodily injury as a direct result of participation in a research project." Section VII(G).
Wake Forest University Research Related Injury Operational Policy and Procedure (last revised 2007)	"It is the position of the Wake Forest University School of Medicine (WFUSM) Institutional Review Board (IRB) that for any research study it approves and determines to be of greater than minimal risk, provisions must be made for the coverage of reasonable costs for the necessary treatment for illnesses, adverse events or injuries that results from medications, devices, interventions, procedures, or tests that the research study subject would not have been exposed to had he or she not volunteered to participate in the research study." Research Related Injury. "For research studies of greater than minimal risk that are **industry sponsored**, provisions must be made for the coverage of reasonable costs for the necessary treatment for illness, adverse events or injuries that results from medications, devices, interventions, procedures, or tests that the research study subject would not have been exposed to had he or she not volunteered to participate in the research study protocol." Research Related Injury 2. [Bold added] "The Wake Forest University School of Medicine maintains limited liability insurance coverage to provide for the treatment of research related injuries that occur as a result of participation in **non-sponsored** research (e.g. NIH or Departmental)." Research Related Injury 3. [Bold added]

Appendix IV: International and Transnational Requirements for Treatment and Compensation for Research Injuries

Australia Note for Guidance on Good Clinical Practice CPMP/ICH/135/95 (2000) National Statement on Ethical Conduct in Human Research (2007)	Institutions must be satisfied that sponsors of trials have made the indemnity or insurance and compensation arrangements required by *CPMP/ICH Note for Guidance on Good Clinical Practice* [recommending that care be provided, and that compensation be provided in accordance with local requirements], *ISO 14155 Clinical Investigation of Medical Devices* [requiring disclosure of provisions made for compensation] and the TGA. Art. 3.3.24 (2007).
Austria Austrian Drug Law (2009) Federal Act on Medical Devices (as amended 2011)	The sponsor must have personal injury insurance to cover any injuries that may be caused to life or health of the subject by the tests carried out in the clinical trial and for which the clinical investigator too would be liable if at fault. Art. 47(1) (2011).
Belgium Law Relating to Experimentation on Humans, Chapter XVII (Responsibility and Insurance) (2004)	Before commencing the experiment, the sponsor shall enter into an insurance contract which covers this liability, and the liability of every person intervening in the trial, irrespective of the nature of the affiliation between the intervening person, the sponsor and the subject. Art. 29.2.
Bosnia/Herzegovina Medicinal Products and Medical Devices Act (2008)	Prior to commencement of the testing, the legal entity performing the clinical testing of the medical device and the sponsor of the clinical testing shall insure their liability against any possible damages which might be caused to the participant or participants in the clinical trial. Art. 116.
Brazil Rules on Research Involving Human Subjects (Res. CNS 196/96 and others) (2003)	The researcher, the sponsor and the institution must assume full responsibility for providing comprehensive care to the research subjects, as regards complications and injury resulting from foreseen risks. Art. V.5.
Bulgaria Law on the Medicinal Products in Human Medicine (2007, amended through 2011)	The contracting authority and the principal researcher shall have insurance covering their liability for property or non-property damages to the study subjects caused in or on the occasion of the conduct of the clinical test. Art. 91.
Canada Tri-Council Policy Statement: Ethical Conduct for Research Involving Humans (2010)	The information generally required for informed consent includes: … (j) information about any payments, including incentives for participants, reimbursement for participation-related expenses and compensation for injury. Art. 3.2.
China Good Clinical Practice (Board Order No. 3) (2003)	The sponsor should provide insurance for the subjects participating in clinical trials so that injured subjects do not bear the cost of treatment and so they may receive corresponding economic compensation. Art. 43.
CIOMS[*] Council for International Organizations of Medical Sciences (CIOMS), International Ethical Guidelines for Biomedical Research Involving Human Subjects (2002)	Investigators should ensure that research subjects who suffer injury as a result of their participation are entitled to free medical treatment for such injury and to such financial or other assistance as would compensate them equitably for any resultant impairment, disability or handicap. Guideline 19.
Council of Europe[+] Convention for the Protection of Human Rights and Dignity of the Human Being with regard to the Application of Biology and Medicine: Convention on Human Rights and Biomedicine (1997) Additional Protocol to the Convention on Human Rights and Biomedicine concerning Biomedical Research, Article 13, CETS No. 195 (2005)	The person who has suffered damage as a result of participation in research shall be entitled to fair compensation according to the conditions and procedures prescribed by law. Art. 31 (2005).

[*] International standard setting body
[+] Supranational European body

Appendix IV: International and Transnational Requirements for Treatment
and Compensation for Research Injuries

Croatia Clinical Trials and Good Clinical Practice (2003)	The Central Ethics Committee, in the approval process of clinical testing, will determine: the existence of insurance compensation in case of injury, death or treatment of subjects that is related to clinical trials. Art. 12.
Declaration of Helsinki - World Medical Association[*] 59th WMA General Assembly Seoul, Korea (2008)	The protocol should include information regarding ... provisions for treating and/or compensating subjects who are harmed as a consequence of participation in the research study. Section B.14.
Denmark The Danish Liability for Damages Act (2005, amended 2006 and 2007) Danish Act on the Right to Complain and Receive Compensation within the Health Service (2009)	Compensation shall be paid pursuant to the provisions of this Part to patients or the bereaved families of patients who suffer injury in Denmark in connection with examination, treatment or the like carried out... Individuals taking part in biomedical trials that do not form part of the diagnosis or treatment of their illness shall have the same status as patients. Part 3 §§19(1), (2) (2009).
Estonia Medicinal Products Act (2004; amended most recently in 2010) Conditions and Procedure for Conducting Clinical Trials of Medicinal Products, Regulation No. 23 of the Minister of Social Affairs (2005)	The sponsor of a clinical trial of a medicinal product shall guarantee the trial subjects health insurance protection in the event of damage to health related to the trial. §90(9) (2004).
European Union[+] Directive 2001/20/EC of the European Parliament and of the Council (2001)	A clinical trial may be undertaken only if, in particular: ... (f) provision has been made for insurance or indemnity to cover the liability of the investigator and sponsor. Art. 3.2(f).
Finland Patient Injuries Act (586/1986; amendments up to 1100/2005 included) (amendments through 2005)	The Patient Injuries Board is responsible for issuing recommendations for decisions on individual claims at the request of a claimant... and for issuing, when requested by a court or one of the parties involved, statements on compensation claims which are being processed in court. Section 11a(1).
France Biomedical Research (Loi Huriet-Sérusclat), Articles L1121-1 to L1126-7 (2004)	Requires that entities both private (e.g. industry, individuals, or private institutions) and public (government or nongovernmental institutions) serving as sponsors must obtain insurance to cover the costs for all damages or injuries arising from the performance of the trial.
Germany Medicinal Products Act (The Drug Law) (2009)	The clinical trial of a medicinal product may only be conducted on human beings if and as long as: ... 8. in the event that a person is killed or a person's body or health is injured during the course of the clinical trial, an insurance policy which provides benefits, even when no one else is liable for the damage, exists .... Section 40(1).
Hungary Act XCV of 2005 on Medicinal Products for Human Use and on the Amendment of Other Regulations Related to Medicinal Products (2005)	The sponsor of a clinical trial shall obtain sufficient liability insurance coverage for any damages that may occur in connection with the clinical trial from an insurance company that is established or has a branch in any Member State of the European Economic Area ... . The liability insurance policy shall afford sufficient cover for any and all potential claims for damage in connection with the clinical trial. Section 3(5).

[*] International standard setting body
[+] Supranational European body

continued

Iceland Regulation on Clinical Trials of Medicinal Products in Humans No. 443 (2004)	Subjects participating in a clinical trial of a medicinal product must be sufficiently insured against conceivable damage to their health resulting from the trial. The principal investigator or, as the case may be, the investigator shall be responsible for ensuring satisfactory insurance coverage. Art. 5.
ICH Harmonised Tripartite Guideline: Guideline For Good Clinical Practice E6(R1)[*] International Conference on Harmonisation of Technical Requirements for Registration of Pharmaceuticals for Human Use (1996)	During and following a subject's participation in a trial, the investigator/institution should ensure that adequate medical care is provided to a subject for any adverse events … related to the trial. Art. 4.3.2. The sponsor's policies and procedures should address the costs of treatment of trial subjects in the event of trial-related injuries in accordance with the applicable regulatory requirement(s). Art. 5.8.2.
India Ethical Guidelines for Biomedical Research on Human Participants (2006)	The sponsor whether a pharmaceutical company, a government, or an institution, should agree, before the research begins, in the a priori agreement to provide compensation for any physical or psychological injury for which participants are entitled or agree to provide insurance coverage for an unforeseen injury whenever possible. Chapter III, Section VI.
Institute of Medicine[*] Responsible Research: A Systems Approach to Protecting Research Participants (2002)	Organizations conducting research should compensate any research participant who is injured as a direct result of participating in research, without regard to fault. Compensation should include at least the costs of medical care and rehabilitation, and accrediting bodies should include such compensation as a requirement of accreditation. Recommendation 6.8.
Ireland Statutory Instruments, S.I. No. 190 of 2004, European Communities (Clinical Trials on Medicinal Products for Human Use) (2004)	In preparing its opinion, the ethics committee shall consider, in particular, the following matters: … (k) the provision made for indemnity or compensation in the event of injury or death attributable to the clinical trial; (l) any insurance or indemnity to cover the liability of the investigator and sponsor…. Part 3, Section 13(6).
Israel Guidelines for Clinical Trials in Human Subjects (2006)	A commercial company entering into an agreement with a medical institution and/or Investigator to conduct a clinical trial shall insure its legal liability pursuant to the laws of Israel against claims filed by clinical trial participants and/or third-party claims in connection with the clinical trial, whether during the course of the trial or thereafter. Appendix 2.
Italy Minimum Requirements for Insurance Policies to Protect the Subjects Participating in Clinical Trials of Medicines (2009)	The clinical trial sponsor must submit to the ethics committee a certificate of insurance. … The insurance policy must provide coverage to specific compensation for injuries. Arts. 1.1, 1.2.
Japan Pharmaceutical Administration and Regulations in Japan (2011)	Insurance coverage and other measures required for compensation in cases of trial-related injury must be undertaken beforehand. Chapter 3, page 101.

[*] International standard setting body
[+] Supranational European body

Appendix IV: International and Transnational Requirements for Treatment and Compensation for Research Injuries

Latvia

Cabinet Regulation No. 289: Regulations on Conducting Clinical Trials and Non-Interventional Studies and Labeling of Investigational Medicinal Products, and Procedure for Conducting Inspections on Compliance with the Requirements of Good Clinical Practice (2010)

The sponsor shall ensure that provisions have been made for insurance and indemnity to cover the liability of the investigator and sponsor. The sponsor is not responsible for a deliberate or accidental injury to a subject caused by the investigator or other individuals involved in the clinical trial. Art. 22.

Lithuania

Principal Investigators and Biomedical Research Customers in Civil Liability Insurance Rules (2000)

Liability Compulsory Insurance Regulations govern the principal investigators and biomedical research customers' compulsory insurance contract, the conditions of the contract, the parties - the insurer and the insured, the pre-contractual and contractual rights and obligations, and the insured person's rights and obligations. Art. I.1.

Macedonia (Republic of)

On the Manner and the Procedure for Clinical Trials on Medicinal Products and the Documentation (2009)

The application for granting approval for the clinical trial of the medicinal product referred to in Paragraph 1 of this Article shall contain: ... 14. evidence that the trial subjects had been insured by the applicant in case of occurrence of damage to the investigator's health (damage or death of the trial subject). Section II(7), Art. 24.

Netherlands

Rules for Compulsory Insurance in Medical Research Involving Human Subjects (2003)

Medical Research Involving Human Subjects Act (WMO) (2006)

The trial shall not be conducted unless at the time of its commencement a contract of insurance has been entered into covering losses due to death or injury resulting from the trial. Such insurance need not cover injury which is inevitable or almost inevitable, given the nature of the trial. Section 7.1.

New Zealand

Health Research Council: Guidelines on Ethics in Health Research (2002, revised 2005)

Injury Prevention, Rehabilitation, and Compensation Act (2001, amended 2007)

The Injury Prevention, Rehabilitation, and Compensation Act 2001 (IPRC Act), provides cover for treatment injuries caused as part of a clinical trial where an accredited ethics committee has approved the trial and is satisfied that the trial was not to be conducted principally for the benefit of the manufacturer or distributor of the medicine or item being trialed. Art. 5.6 (2005).

Philippines

Philippine Council for Health Research and Development, National Guidelines for Biomedical/Behavioral Research (2000)

National Ethical Guidelines for Health Research (2006)

The investigator shall provide the following information to the potential subject, using language that can be understood: ... w. That treatment will be provided free of charge for specified types of research-related injury or for complications associated with the research, the nature and duration of such care, the name of the organization or individual that will provide the treatment, and whether there is any uncertainty regarding funding of such treatment; x. In what way, and by what organization the subject or the subject's family or dependents will be compensated for disability or death resulting from such injury (or, when indicated, that there are no plans to provide such compensation). Art. I.2 (2006).

Poland

Pharmaceutical Law of 6 September 2001 (amended through 2008)

The Bioethics Committee shall consider, in particular: ... (9) the level of the indemnity or compensation in the event of injury or death attributable to participation in the clinical trial. Art. 37r(2).

Russia

On Medicinal Products, Federal Law No. 86-FZ (2006)

The legal basis for conducting clinical trials of medicinal products consists of the following documents: 5) civil liability insurance of persons carrying out clinical trials of the drug. Article 38.1

* International standard setting body
+ Supranational European body

continued

Singapore Medicines (Clinical Trials) Regulations (2000)	The subject is entitled to a full and reasonable explanation of the following: … (j) any compensation and treatment available to the subject in the event of injury arising from participation in the clinical trial. Art. 14(1).
Slovakia On Healthcare, Healthcare-Related Services and on the of Certain Laws, Act No. 576/2004 Coll. (2004)	The instructions prior to informed approval must contain information on … (h) measures intended for ensuring adequate compensation in the case of damage of the health of the research participant in connection with his/her participation in this research. Art. 27(2).
South Africa Guidelines on Ethics for Medical Research: General Principles (2002)	In the event of significant injury, the participant should be entitled to receive compensation regardless of whether or not there was negligence or legal liability on any other basis. Art. 10.6.2.2.
South Korea Guideline for Korean Good Clinical Practice (2000)	If required by the applicable regulatory requirement(s), the sponsor shall provide insurance or shall indemnify (legal and financial coverage) the investigator/the institution against claims arising from the trial, except for claims that arise from malpractice and/or negligence. Art. 32.
Spain Royal Decree 223/2004 of 6 February Regulating Clinical Trials with Medicinal Products (2004) Law 14/2007, of 3 July, on Biomedical Research (2007)	A clinical trial with investigational medicinal products may only be undertaken if insurance or other financial cover has previously been taken out to include any trial-related injury or loss occurring to the person in whom it is to be conducted…. Art. 8(1) (2004).
Switzerland SR 812.214.2 Ordinance on Clinical Trials of Therapeutic Products (2010)	The developer is liable for injuries to a research subject in the context of a clinical trial through insurance to cover liability. Arts. 7(1), 7(2).
Taiwan Human Research Ethics Policy Guidelines (2007) Medical Care Act (2009)	In the case that violation by juridical persons in medical care results in damage or injury, the offender shall be responsible for compensation. Art. 112 (2009).
Uganda National Guidelines for Research Involving Humans as Research Participants - Uganda National Council for Science and Technology (2007)	Research participants shall be entitled to compensation when the injury is classified as 'Probably' or 'Definitely' related to their participation in the research project. Sponsors shall ensure that research participants who suffer injury as a result of their participation in the research project are entitled to free medical treatment for such injury and to such financial or other assistance as would compensate them equitably for any resultant impairment, disability or handicap. Art. 7.5.
Ukraine On Medicines, No. 3323-VI (2011)	The sponsor of the clinical trial shall, prior to its performance, make an Agreement on Insurance of Health and Life of the Patient (the Volunteer), according to the procedure envisaged by the legislation. Art. 8.
United Kingdom The Medicines for Human Use (Clinical Trials) Regulations, 2004 No. 1031 (2004)	In preparing its opinion, the committee shall consider, in particular, the following matters … (i) provision for indemnity or compensation in the event of injury or death attributable to the clinical trial. Art. 15(5)(i).

* International standard setting body
+ Supranational European body

V

Appendix V: International Research Panel

Members

Amy Gutmann, Ph.D., Chair*
President and Christopher H. Browne Distinguished Professor of Political Science, University of Pennsylvania

John D. Arras, Ph.D.*
Porterfield Professor of Biomedical Ethics
Professor of Philosophy
University of Virginia

Julius Ecuru, B.Sc., M.Sc., Dip. IRE.
Assistant Executive Secretary
Uganda National Council for Science and Technology, Kampala-Uganda

Christine Grady, R.N., Ph.D.*
Acting Chief of the Department of Bioethics
National Institutes of Health
Clinical Center

Dirceu Bartolomeu Greco, M.D., Ph.D.
Professor, Internal Medicine
School of Medicine
Federal University of Minas Gerais
Belo Horizonte, Brazil

Unni Karunakara, M.B., B.S., Dr.PH.
Assistant Clinical Professor, Heilbrunn Department of Population & Family Health
Mailman School of Public Health
Columbia University

Nandini K. Kumar, M.B.B.S., D.C.P., M.H.Sc.
Former Deputy Director General Sr. Grade (Scientist F)
Investigator NIH project on Bioethics
National Institute of Epidemiology, India

Sergio G. Litewk a, M.D., M.P.H.
Research Assistant Professor
Director International Programs
CITI Program Latin American Coordinator
University of Miami

Luis Manuel López Dávila, M.D., M.Soc. Sc., M.A.
Researcher of the Year 2009
Medical School Professor, and Committee on Bioethics in Health Research Staff Member
Universidad de San Carlos de Guatemala
President and Founder of Oxlajuj N'oj Foundation

Adel Mahmoud, M.D., Ph.D.
Professor in Molecular Biology and Public Policy
Woodrow Wilson School of Public and International Affairs
Princeton University

Nelson L. Michael, M.D., Ph.D. *
Colonel, Medical Corps, U.S. Army
Director, Division of Retrovirology
Walter Reed Army Institute of Research
U.S. Military HIV Research Program

Peter Piot, M.D., Ph.D.
Director and Professor of Global Health
London School of Hygiene & Tropical Medicine, UK

Huanming Yang, Ph.D.
Professor & President
BGI (Beijing Genomics Institute), China

Boris Yudin, Ph.D.
Corresponding Member of the Russian Academy of Sciences, Professor of Philosophy
M. Lomonosov's Moscow State University

*Commission member

Appendix VI: Guest Speakers

Ronald Bayer, Ph.D.
Professor and Co-Chair, Center for the History and Ethics of Public Health, Mailman School of Public Health, Columbia University

David A. Borasky, Jr., M.P.H., C.I.P.
IRB Manager, Office of Research Protection, RTI International

Dan W. Brock, Ph.D.
Frances Glessner Lee Professor of Medical Ethics, Department of Global Health and Social Medicine; Director, Division of Medical Ethics, Harvard Medical School

Robert M. Califf, M.D.
Vice Chancellor for Clinical Research, Duke University; Professor of Medicine, Duke University Medical Center; Director, Duke Translational Medicine Institute

Connie Celum, M.D., M.P.H.
Professor of Medicine; Professor of Global Health; Adjunct Professor of Epidemiology; Director, International Clinical Research Center, Department of Global Health, University of Washington

Lawrence Corey, M.D.
President and Director, Fred Hutchinson Cancer Research Center; Principal Investigator, HIV Vaccine Trials Network; Professor, Laboratory Medicine and Medicine, University of Washington

Francis P. Crawley
Executive Director, Good Clinical Practice Alliance – Europe (GCPA)

Ezekiel J. Emanuel, M.D., Ph.D.
Diane v.S. Levy and Robert M. Levy University Professor and Vice Provost for Global Initiatives, Center for Bioethics, University of Pennsylvania

Kenneth R. Feinberg, J.D.
Administrator of the Gulf Coast Claims Facility and Special Master of the September 11 Victim Compensation Fund (2001-2004)

Dafna Feinholz, Ph.D.
Chief of the Bioethics Section, Division of Ethics of Science and Technology, Sector for Social and Human Services, United Nations Educational, Scientific and Cultural Organization (UNESCO)

Joseph J. Fins, M.D., F.A.C.P.
E. William Davis, M.D., Jr. Professor of Medical Ethics; Chief, Division of Medical Ethics; Professor of Medicine, Professor of Public Health, and Professor of Medicine in Psychiatry, Weill Cornell Medical College

Jeffrey K. Francer
Assistant General Counsel, PhRMA

Roger I. Glass, M.D., Ph.D.
Director, Fogarty International Center; Associate Director for International Research, National Institutes of Health

Susan E. Lederer, Ph.D.
Robert Turell Professor of History of Medicine and Bioethics; Chair, Medical History and Bioethics University of Wisconsin School of Medicine and Public Health

Appendix VI: Guest Speakers

Sergio Litewka, M.D., M.P.H.
International Programs Director;
Research Assistant Professor,
University of Miami Ethics Programs

Murray M. Lumpkin, M.D., M.Sc.
Deputy Commissioner for
International Programs, U.S. Food
and Drug Administration (FDA)

Ruth Macklin, Ph.D.
Professor of Bioethics, Department of
Epidemiology and Population Health,
Albert Einstein College of Medicine

Russell M. Medford, M.D., Ph.D.
Chairman and President of
Salutria Pharmaceuticals

Jerry Menikoff, M.D., J.D.
Director, Office for Human Research
Protections, U.S. Department of Health
& Human Services

Eric M. Meslin, Ph.D.
Director, Indiana University Center for
Bioethics; Dean (Bioethics), IU School
of Medicine; Professor of Medicine,
Medical and Molecular Genetics,
Public Health, and Philosophy; Director,
Indiana University-Moi University
Academic Research Ethics Partnership

Karen E. Moe
Director & Assistant Vice Provost
for Research, University of Washington
Human Subjects Division

Linda Nielsen, J.D.
Vice President, European Group on
Ethics in Science and New Technologies
(EGE); Professor of Global Law
and Governance, University of
Copenhagen, Denmark

Maurizio Salvi, Ph.D.
Policy Advisor (Ethics) to the President of
the European Commission; Head of the
EGE Secretariat; Member of the Bureau
of European Policy Advisors (BEPA)

Robert Temple, M.D.
Deputy Center Director for Clinical
Science, Center for Drug Evaluation
and Research; Acting Director, Office
of Drug Evaluation I, U.S. Food and
Drug Administration (FDA)

Carletta Tilousi
Member, Havasupai Tribal Council
Havasupai Tribe

Johannes J. M. van Delden, M.D., Ph.D.
President, Council for International
Organizations of Medical Sciences (CIOMS);
Professor of Medical Ethics, Utrecht
University Medical Center

Mitchell Warren
Executive Director, AVAC: Global
Advocacy for HIV Prevention

Daniel Wikler, Ph.D.
Mary B. Saltonstall Professor of Population
Ethics, Professor of Ethics and Population
Health, Department of Global Health and
Population, Harvard University

John R. Williams, Ph.D.
Consultant, World Medical Association
(WMA); Adjunct Professor, Department
of Medicine, University of Ottawa;
Adjunct Research Professor, Department of
Philosophy, Carleton University (Ottawa)

www.ingramcontent.com/pod-product-compliance
Lightning Source LLC
Chambersburg PA
CBHW080243180526
45167CB00006B/2391